D0025066

THE STRATEGIC PETROLEUM RESERVE

Planning, Implementation, and Analysis

David Leo Weimer

Contributions in Economics and Economic History, Number 48

GREENWOOD PRESS
WESTPORT, CONNECTICUT • LONDON, ENGLAND

Library of Congress Cataloging in Publication Data

Weimer, David Leo.
The strategic petroleum reserve.

(Contributions in economics and economic history,
ISSN 0084-9235 ; no. 48)
Bibliography: p.
Includes index.
1. Petroleum industry and trade—Government policy—
United States. 2. Strategic materials—United States.
I. Title. II. Series.
HD9566.W44 333.8′23211′0973 82-6184
ISBN 0-313-23404-3 (lib. bdg.) AACR2

Library of Congress Catalog Card Number: 82-6184
ISBN: 0-313-23404-3
ISSN: 0084-9235

First published in 1982

Greenwood Press
A division of Congressional Information Service, Inc.
88 Post Road West, Westport, Connecticut 06881

Printed in the United States of America

10 9 8 7 6 5 4 3 2 1

To Bessie and the memory of Roy

Contents

Figures and Tables ix

Preface xi

Acknowledgments xiii

Abbreviations xv

Part I ANALYSIS OF SPR HISTORY

1. Introduction: Origins of the Strategic Petroleum Reserve Program 3
2. Getting Started Under FEA: Strategic Planning 22
3. Debacle Under DOE: Sources of Implementation Failure 39
4. The Will and Wherewithal to Fill 63

Part II HISTORY OF SPR ANALYSIS

5. Basic Analytical Considerations for SPR Size Studies 83
6. The Size Issue and OMB: Analysis in the Bureaucracy 111
7. The Role of the Private Sector in Strategic Stockpiling 141
8. Distribution and Drawdown 164

Part III CONCLUSION

9. Perspectives on SPR History 185

Appendix: The SPR Provisions of EPCA 199

Bibliography 209

Index 223

Figures and Tables

Figures

1. SPR Plan: Major Implementation Actions 34
2. An Underlying World Oil Price Trajectory 88
3. Basic World Oil Price Assumptions 89
4. Microeconomic Costs of Acquisitions and Disruptions 91
5. Oil Price Shocks in the Aggregate Supply and Demand Framework 95
6. Decision Tree for Simple Stockpiling Problems 100
7. Numerical Example of Decision Tree Analysis 101

Tables

1. Summary Analysis of Storage Alternatives in the SPR Plan (December 1976) 28
2. SPR Phase I Facilities (248 million barrels of capacity) 48
3. Strategic Petroleum Reserve Appropriations 58
4. Review of Major Analytical Assumptions Made in SPR Evaluations 104
5. Net Economic Benefits of the SPR 132

Preface

Twice during the last decade the United States suffered severe economic losses from disruptions of the world oil market. The 1973 Arab oil embargo and the reduction in oil production by Iran during its revolution precipitated sharp rises in the world price of oil that reduced economic activity and increased inflation in the industrialized nations of the West. The volatile political climate of the Middle East, the origin of a significant fraction of the oil imported by the United States and its allies, makes disruptions of the world oil market over the next decade not just possible but highly probable. The release of petroleum stockpiles during market disruptions can substantially reduce economic losses by moderating price rises and reducing the transfers of wealth from oil importers to oil exporters.

Before stockpiles can be released, however, they must be accumulated. High carrying costs and fear of expropriations by government discourage private firms from holding petroleum stocks larger than needed to maintain minimum flows through pipeline systems and to accommodate seasonal shifts in demand. Japan and several European nations have built up strategic reserves of oil by mandating increased storage by their petroleum industries. Although Congress has authorized a similar approach for the United States, the Ford,

Carter, and Reagan administrations have pursued instead the development of crude oil stockpiles fully owned and controlled by the federal government. These reserves of crude oil, which are stored in salt formations in Louisiana and Texas, constitute the strategic petroleum reserve (SPR) of the United States.

Rarely has a program concept received such widespread support. Presidents and prominent congressmen of both parties, a succession of cabinet members, petroleum industry spokesmen, and national security and energy policy experts both inside and outside of government have advocated the creation of a large SPR. The only significant opponents have been officials in the Office of Management and Budget and a former governor of Louisiana. Despite the preponderance of political support for the SPR, stockpile goals have not been met. In 1975 Congress set as a goal the storage of 500 million barrels of crude oil and petroleum products by the end of 1982. Two years later, with congressional approval, the Carter administration advanced the 500 million barrel goal to the end of 1980. By the end of 1980, only about 100 million barrels were actually in storage and the anticipated completion date for the 500 million barrels of storage capacity had slipped to the end of 1985. Why have the SPR goals not been met?

The first part of this study seeks an answer to this question in the history of the implementation of the SPR program. Although it is hoped that the study of SPR implementation will contribute to our growing understanding of why some programs fail and others succeed, the primary intellectual contribution is intended to arise in the second part of the study which focuses on the assumptions, conclusions, and organizational influences inherent in the analyses that have shaped strategic petroleum reserve policies. The literature of the policy sciences abounds with many studies of programs but few studies of the analysis behind them. The first two parts of the study are intended to bridge this gap. The third part draws on the first two to suggest more appropriate stockpiling policies for the United States.

Acknowledgments

My interest in the strategic petroleum reserve developed while I was a visiting economist with the Office of Energy Security in the Department of Energy. I was fortunate to serve under Lucian Pugliaresi. Not only did he make my visit an enjoyable and professionally valuable experience, he encouraged me to write this book. Without his encouragement and support and that of Glen Sweetnam, Jerry Blankenship, Mike Barron, and Steve Minihan, I would never have completed it.

I am very grateful to all the Strategic Petroleum Reserve Office personnel, past and present, who assisted me in my research. A special thanks to Carlyle Hystad and Howard Borgstrom, who were particularly generous with their time and to Robert Davies, Joseph DeLuca, Jay Brill, and Harry Jones, who shared their insights as program directors. Others in the Department of Energy, the Office of Management and Budget, and the Council of Economic Advisors deserve my thanks.

I am in debt to all those who commented on various parts of the manuscript. I benefited from the extensive comments of Eric Hanushek, Nancy Wentzler, Marshall Hoyler, and Hank Jenkins-Smith. Bruce Jacobs, Anthony Koo, George Horwich, Connie Nelson, Kathy Abbott, David Gustafson, Christopher Flavin, Aaron Wildavsky, Eugene Bardach, William Hogan, Stephen Wasby, and Michael Wolkoff also offered helpful advice or comments. Needless to say, I bear the sole burden for remaining errors.

My thanks to Donna French, Christopher Zehren, Scott Newman, and Hank Jenkins-Smith for their valuable assistance in preparation of the final manuscript.

Abbreviations

BSA	basic sales agreement
CBO	Congressional Budget Office
CEA	Council of Economic Advisors, Executive Office of the President
DFSC	Defense Fuel Supply Center, Department of Defense
DOE	Department of Energy (subsumed ERDA, FEA, and FPC)
DRI	Data Resources, Inc.
EBV	Erdoelbevorratungsverband
EIS	environmental impact statement
EPA	Environmental Protection Agency
EPAA	Emergency Petroleum Allocation Act of 1973
EPCA	Energy Policy and Conservation Act of 1975
ERA	Economic Regulatory Administration
ERDA	Energy Research and Development Administration
ESR	Early Storage Reserve
FEA	Federal Energy Administration (replaced FEO)
FEO	Federal Energy Office, Executive Office of the President

FPC	Federal Power Commission
FY	Fiscal Year
GAO	General Accounting Office
GNP	gross national product
IEA	International Energy Agency
IPR	industrial petroleum reserve
NEP	National Energy Plan
NOAA	National Oceanic and Atmospheric Administration
NPR	naval petroleum reserve
OAPEC	Organization of Arab Petroleum Exporting Countries
OMB	Office of Management and Budget, Executive Office of the President
OPEC	Organization of Petroleum Exporting Countries
PE	Office of the Assistant Secretary for Policy and Evaluation, DOE
PIES	Project Independence Evaluation System
PL	Public Law
PPA	Office of Policy, Planning and Analysis, DOE (formerly PE)
RPR	regional petroleum reserve
SEB	source evaluation board
SPR	strategic petroleum reserve

PART I

ANALYSIS OF SPR HISTORY

1

Introduction: Origins of the Strategic Petroleum Reserve Program

Since 1976 the U.S. Government has been committed to stockpiling large quantities of crude oil in underground salt dome caverns through the strategic petroleum reserve (SPR) program. Filling holes in the ground with oil may seem like a strange way to deal with the "energy crisis." The value of having petroleum stockpiles becomes apparent, however, if we recognize that from a public policy perspective the real "energy crisis" is the vulnerability of the United States and its allies to disruptions of the world oil market. Although implementation has been problematical, the development of large and readily available petroleum stocks is probably the single most important element of our national energy policy.

Access to adequate supplies of petroleum is an important part of the economic and military security of modern industrialized nations. Steep price increases, due to either purposeful or accidental reductions in the volume of oil offered for sale in the world market, can idle workers and capital, interfere with preferred consumption patterns, slow investment, accelerate inflation, and result in immense transfers of wealth from oil consumers to oil producers. Interference by governments in their own domestic markets or in the world market can cause or exacerbate economic losses by preventing scarce supplies from being allocated by the market to those who will derive the highest value from their use in production and consumption. Access to oil takes on political and military dimensions when nations attempt to secure concessions by such actions as embargoing exports or blocking critical transportation routes. Restrictions on access, actual or feared, can be important factors in the initiation, conduct, and resolution of wars.[1]

Vulnerability and Dependence

It is important to distinguish between dependence and vulnerability. The United States is dependent on foreign oil. In 1980 net imports of crude oil and petroleum products averaged 6.2 million barrels per day, or about 36 percent of consumption.[2] If the volume of oil supplied to the world market was not subject to sudden change, dependence on foreign oil would not be a serious problem. In general it is economically better to import additional units of any product as long as the price is less than the cost of domestically producing additional units. Because of balance of payments effects and the fact that import levels can affect the world price, the social cost to the United States of the marginal barrel of oil imported may be higher than the market price.[3] If foreign supplies were secure, a tariff on oil imports equal to the difference between the social costs and world price would be economically efficient. Even with the tariff, however, imports might remain at a high level. Furthermore, achieving independence would not eliminate U.S. vulnerability to disruptions of the world oil market as long as allies and trading partners continued to be net importers of oil. While most reductions in dependence contribute somewhat to reduced vulnerability by reducing the size of wealth transfers to foreign oil producers that result from price shocks, some reductions in import levels, such as those achieved through behavioral changes (driving less, lowering thermostat settings, and so on), may actually increase vulnerability by using up the relatively easy ways to quickly reduce consumption during supply emergencies. The point is not that such conservation is necessarily bad, but rather that a clear distinction needs to be made between dependence and vulnerability; the United States may greatly decrease its dependence on imported oil while remaining highly vulnerable to disruptions of the world oil market.

The most important policy change with respect to U.S. dependence on foreign oil was taken in February 1981 by eliminating the price controls on domestically produced oil that had been in effect since August of 1971. The entitlements system, which was intended to equalize the access of refiners to lower priced domestic oil, had the effect of subsidizing imports. Beyond removing this subsidy and instituting a tariff on oil imports, some governmental role in encouraging investments in conservation and supporting long-term research into alternative energy sources may be justified.[4] Reductions in dependence through such government actions will reduce but not eliminate, for at least the next ten to twenty years, U.S. vulnerability to disruptions of the world market.

One approach for reducing vulnerability, increasing the security of supply to the world oil market, will prove extremely difficult to implement. Encouraging production of energy resources in relatively stable regions of the world offers some hope but is ultimately limited by the oil abundance of the Middle East. Because it enjoys huge reserves that can be produced at very low costs, the Middle East will continue to be a major source of world energy for the foreseeable future. Significant improvements in the security of supply can therefore only be made by increasing the political stability of the Middle East. The sources of instability in the region are numerous: religious and ethnic differences, historical animosities, wealth disparities, nationalism, accelerated modernization, and the threat of Soviet intervention. This political complexity makes it difficult for the United States to determine, let alone execute, a foreign policy that clearly increases stability. For example, the United States believed it was contributing to regional stability by supporting Mohammad Reza Pahlavi, the Shah of Iran. An increased military presence may reduce the dangers of Soviet intervention but at the same time contribute to internal disorder in the Persian Gulf states that provide U.S. bases. Even peaceful resolution of the Arab-Israeli conflict, as desirable as it is for other reasons, might lead to greater conflict among Arab states no longer facing a common enemy.

A more certain approach for reducing vulnerability is to develop the capability for delivering stocks of oil and other energy supplies to the world market when there is a temporary reduction in the volume of oil normally supplied. Increased capabilities for fuel switching by oil users would permit available stocks of coal and natural gas to "back-out" oil imports. The development of "surge" production capabilities at existing oil fields provides in situ storage of reserve stocks. In fact, production prorating by state regulatory agencies in Texas and Louisiana provided surge capacity that was used to offset the world oil shortfall resulting from the Suez crisis in 1956. Since 1972, when Texas and Louisiana prorating reached 100 percent of the maximum efficient rate of production, there has been little surge capacity available in the United States. Oil fields could be once again shut-in, but a less costly and more flexible way to make available surge stocks is the stockpiling of oil in man-made facilities.

Stockpiling as Policy

Stockpiling as a governmental policy has many precedents. Genesis tells us how the pharaoh prospered during the seven bad years because he stockpiled grain during the seven good ones, as Joseph had recommended. The primary function of the administrative centers of the Incan Empire appears to have been the stockpiling of agricultural products. In more modern times, the primary function of agricultural stockpiling programs, such as the U.S. Commodity Credit Corporation, has been the protection of farmers during times of plenty rather than the protection of consumers from high prices during times of scarcity. The reliance of modern industry on worldwide sources of raw materials that might become unavailable during times of war has prompted nations to store nonagricultural commodities.

In 1939 the United States began the strategic and critical materials stockpiling program to insure that wartime disruption of imports of raw materials would not stop industrial production.[5] Stockpiling under the program has ranged from the 1946 goal of having adequate stocks to replace lost imports during a five-year conflict to the 1973 goal of having adequate stocks to replace imports lost from Communist and potential combatant countries during a one-year conflict. The Strategic and Critical Materials Stock Piling Revision Act of 1979 (PL96-41), which consolidated the critical materials program with several smaller stockpiling programs, required that stocks of sufficient size to support a three-year war be maintained. As of July 1980 the critical materials stockpile included ninety-three commodities valued at almost $13.5 billion.[6]

Although petroleum is an essential raw material for industrial production, it was not included in the critical materials program. One reason was that, at the time the program was started, the United States was a net exporter of crude oil. It was not until after World War II that the United States became a net importer of petroleum with import levels reaching 20 percent of consumption in 1962 and peaking at almost 48 percent in 1977. Another reason was that the United States had already set aside oil-bearing lands as a naval petroleum reserve (NPR) to serve as in situ stocks for national defense.

Four NPR sites have been established on public lands through executive order.[7] In 1912 the first reserves were set aside at Elk Hills (NPR No. 1) and Buena Vista Hills (NPR No. 2) in California. NPR No. 3, which later became an object of scandal, was established in 1915 at Teapot Dome, Wyoming. NPR No. 4 was established on the Alaskan North Slope in 1933. Lands bearing oil shale were also set aside.

Until 1976 NPR production was limited to levels necessary for maintaining, protecting, and testing the reserves unless Congress authorized increased production for national defense, as it did for NPR No. 1 during the World War II. The Naval Petroleum Reserves Production Act of 1976 (PL94-258) required that the reserves be developed to their maximum efficient rate of production for a period of six years. The resulting production reduces U.S. imports of foreign oil and generates revenues for the Treasury but eliminates what little surge capacity the NPR had provided.

Experience with the NPR suggests several practical problems associated with in situ stockpiling. First, because oil production is governed by the right of capture, protection of reserves requires that sufficient land be set aside to encompass entire reservoirs of oil. This was not the case with NPR No. 2, where the government only owned about 35 percent of the land providing access to the reservoir and therefore has been forced to produce at a maximum rate since the 1920s to avoid losing the oil to commercial producers. Production has also been necessary at NPR No. 1 and No. 3 to prevent loss of oil to commercial producers on nearby land.

Second, sites may not be located near transportation facilities that can accommodate surge production. Development of surge capacity at NPR No. 4 would require that excess capacity be maintained at great expense on the trans-Alaska pipeline. Although NPR No. 4 is an extreme case, the development of in situ surge production capacity generally requires considerable investment in transportation facilities to deliver oil to the commercial distribution system.

Third, equipment and personnel must be kept in readiness for surge production. After the Six Day War in 1967 the secretary of the Navy established an operational readiness requirement for NPR No. 1 of 160,000 barrels per day within sixty days of an order to produce. Because inadequate funds were budgeted for development and maintenance, however, the readiness requirement was never met. In 1976 when commercial production was resumed, a rate of only about 100,000 barrels per day was possible.

Although resolution of these problems is costly, the primary disadvantage of in situ storage is that it removes huge amounts of reserves from current production. For technical reasons it is rarely possible to extract more than one-eighth of a reservoir in a year. Consequently, to provide surge capacity of 3 million barrels per day for one year would require the setting aside of approximately 8.8 billion barrels of reserves, a significant fraction of the current U.S. proven reserves of about 30 billion barrels.[8] Delaying production from set aside reserves is the largest component of cost associated with in situ stockpiling.

There are a variety of methods for holding crude oil once it has been extracted from the ground. Although the petroleum industry has traditionally not held speculative stocks, large volumes of stocks must be held to maintain pipeline flows, accommodate tanker deliveries, and permit efficient production in the face of seasonal demand for petroleum products. Storage in above-ground steel or concrete tanks is by far the most common way stocks are held. In locations where underground salt dome formations are available, it is often economical to create large underground caverns for petroleum storage by leaching processes that dissolve salt with fresh water. The shafts of abandoned mines can also be used for underground storage. When capacity in such facilities is unavailable, firms may lease additional oil tankers or order ones already leased to "slow-steam" until their cargos can be accommodated.

A stockpiling program employing these types of storage could be designed to permit high drawdown rates that would substantially reduce the impact of temporary supply curtailments on the price of oil in the world market. The program would involve three major types of cost: the cost of developing and maintaining storage facilities, the direct costs of purchasing oil, and the economic effects of any price rise caused by the oil purchases. The latter costs could be quite substantial, especially if oil exporting countries retaliate against development of strategic reserves through reductions in production. Drawdowns during disruptions would provide benefits in the form of economic losses avoided. By reducing political pressure for immediate action, drawdowns can provide foreign policy decision makers with a breathing space during which diplomatic initiatives can be attempted and military preparations made. The mere existence of a large stockpile may deter purposeful attempts to disrupt the world market for political and economic gain. By reducing vulnerability to disruptions of the world oil market in these ways, stockpiling greatly reduces the social costs of dependence on foreign oil.

Moves toward a Strategic Petroleum Reserve

The federal government began taking steps to limit the growing dependence on foreign oil during the Eisenhower administration.[9] After several years of unsuccessful jawboning, pressure from the domestic oil and coal industries and the threat of congressional action led the administration to institute a program of voluntary restrictions on oil imports in 1957. When the voluntary restrictions proved ineffective, the administration instituted a system of mandatory import quotas. President Eisenhower acted under authority of the Trade Agreements Extension Act of 1955 (PL84-86), which allows the president to take actions necessary to reduce import levels that "threaten or impair the national security." The quota system remained in effect until 1973, when rising foreign prices, controlled domestic prices, and domestic production of oil at full capacity made it politically unworkable.

In 1971 economists Walter Mead and Philip Sorensen, building on the work of Stephen McDonald, recommended creation of an effective petroleum reserve program as an alternative to the mandatory quota system.[10] They argued for expansion of the naval petroleum reserves and addition of new fields with shut-in production capacity. Based on data provided by the Department of the Interior, they concluded that reserves held in salt domes would also be economically preferable to continuation of the import quota system.

There was growing concern about the increasing vulnerability of the U.S. to oil supply disruptions among a number of analysts in the Departments of Interior and Defense and elsewhere in the executive branch. A succession of energy advisors to the Nixon White House attempted without success to focus the attention of the administration on the vulnerability problem. For example, in 1972 energy advisor James E. Akins argued that drastic measures should be taken to build up inventories in the face of Arab threats. His reward was reassignment.[11] Nevertheless, in December 1972 the Interior Department requested that the National Petroleum Council, an industry advisory group to the department, conduct a study of U.S. vulnerability and options for reducing it. The council was specifically asked to consider the impacts of a 1.5 million barrel per day import shortfall lasting three months and a 3 million barrel per day shortfall lasting six months. In its preliminary report issued in July 1973, the council raised the possibility of protecting the U.S. against the latter shortfall by 1978 through the stockpiling of 540 million barrels of oil in salt dome caverns.[12]

Meanwhile, what little congressional concern there was about the growing U.S. vulnerability was due to the efforts of Senator Henry M. Jackson. As coinitiator and ex officio member of the Senate's National Fuels and Energy Policy Study, he attempted to draw attention to the vulnerability problem and proposed stockpiling as part of the urgently needed solution. In April 1973 he introduced the Petroleum Reserves and Import Policy Act (S.1586). In addition to reformulating the import quota system, it called for creation of a strategic petroleum reserve system consisting of better developed naval petroleum reserves, government-owned stocks held in salt dome caverns, minimum inventory requirements for oil importers, and surge production capacity requirements for domestic oil producers. Senator Jackson explained the need for strategic petroleum reserves at the beginning of hearings on S.1586 before his Committee on Interior and Insular Affairs:

> The uncertainties involved in securing stable sources of Middle East oil are increasingly apparent. In recent months we have seen sabotage of oil installations and the interruption of tanker loadings by guerrilla warfare. We have seen OPEC negotiations founder and U.S. oil interests denounced. Most ominous of all,

> we have seen an increasing tendency to use oil for political blackmail.[13]

Administration, industry, and academic witnesses expressed strong agreement with Senator Jackson's assessment but generally opposed the specific provisions of the bill. Strong objections were raised by most witnesses to the regulatory components of the bill. Prorating of private production on a national basis to provide surge production capacity was viewed least favorably. Industry representatives objected to the provisions mandating minimum inventory levels. Although the development of government-owned stockpiles received the most favorable response, administration witnesses argued that more time was needed for study of stockpiling options.

Events in the Middle East soon made clear the reality of U.S. vulnerability. Egypt attacked Israeli positions across the Suez Canal on October 6, 1973. In the last two weeks of October, the members of the Organization of Arab Petroleum Exporting Countries (OAPEC), led by Saudi Arabia, attempted to use the "oil weapon" in support of Egypt and a favorable settlement of the Arab-Israeli conflict. Shipments of oil to the United States, the Netherlands, and several other countries were embargoed. The OAPEC production rate declined from 20.8 million barrels per day in September to 15.8 million barrels per day in November.[14] The OAPEC production rate was 19 percent lower in the period from November 1973 to February 1974 than it was in September of 1973. Although offset somewhat by increased non-OAPEC production, total world production of oil declined by about 5 percent. By January 1974 U.S. imports were down 2.7 million barrels per day from September of the previous year. Oil prices rose dramatically, almost quadrupling by the time the Arab embargo ended on March 18, 1974. For example, spot market prices for Mideast Light Crude 34 went from $2.70 per barrel in the third quarter of 1973 to $13.00 per barrel in the first quarter of 1974; long-term contract prices for the same oil went from $2.55 in the third quarter of 1973 to $9.60 in the second quarter of 1974.[15] It has been estimated that these price rises reduced the real gross national product of the United States by 7 percent (on an annual basis) during the first quarter of 1974.[16] Many economists consider the price rises to have been the primary cause of the deep recession that extended well into 1975.[17]

President Nixon outlined his proposals for dealing with the Arab embargo and U.S. vulnerability in an address to the nation on November 7, 1973. He called for voluntary conservation, increased production from the naval petroleum reserves, relaxation of environmental standards, deregulation of natural gas prices, and construction of the trans-Alaska pipeline. He also asked for authority to allocate certain petroleum products. (Congress quickly responded in spades with the Emergency Petroleum Allocation Act of 1973 [PL93-159], which applied to all petroleum products and crude oil as well.) He concluded by announcing Project Independence, a national effort to achieve energy self-sufficiency by the end of the decade.

Project Independence was actually launched in March 1974 with an interagency study group eventually led by the newly created Federal Energy Administration (FEA). The Project Independence Report, which was issued in November 1974, recognized that reduction of U.S. vulnerability to oil market disruptions was a more appropriate goal than energy self-sufficiency.[18] Drawing heavily on analysis conducted by the National Petroleum Council, the report considered stockpiling as an option for reducing vulnerability. It suggested that stockpile sizes of a billion barrels or larger might be justified on economic grounds.

The Project Independence Report did not greatly influence the proposals for energy legislation developed by the Ford administration in December 1974. Rather than thoroughly analyzing alternative policies for reducing vulnerability, it had emphasized the modeling of domestic energy supplies and demands for the prediction of import levels. Many of the senior administration officials who were involved in the preparation of the legislative proposals were skeptical of the validity of the models employed and felt the report implied the need for an excessively large role for the government in the energy sector.[19] The resulting legislative package had a stronger market orientation than the Project Independence Report. Although stockpiling was included in the package, Thomas H. Tietenberg suggests that it was viewed as a residual element:

> The strategic reserve had been strongly supported in the Project Independence Report as a very cost-effective way to reduce import vulnerability. In fact, it appeared cheaper than a good number of other domestic strategies. The decision process, however, never performed this trade-off analysis. Instead the

> level of the civilian strategic reserve was established after the rest of the program had been decided upon.[20]

Of all the proposals, however, the strategic petroleum reserve received the most favorable congressional response.

Passage of the Energy Policy and Conservation Act

In January 1975 President Ford sent his proposed Energy Independence Act of 1975 to Congress. It contained thirteen titles. Title I called for full development and production from the naval petroleum reserves and the creation of a military strategic petroleum reserve of 300 million barrels. Title II called for the creation of a domestic strategic petroleum reserve of up to one billion barrels. Other titles called for the deregulation of natural gas, extension of authority to require power plants to use coal rather than oil or natural gas, energy efficiency labeling of appliances and automobiles, thermal efficiency standards for new buildings, revisions of the Clean Air Act, subsidies to encourage low-income families to insulate their dwellings, increased authority for imposing tariffs and quotas on imported oil, standby authority to regulate the allocation of energy resources, development of a national plan for siting energy facilities, and limitation on the delays utilities face in passing on price increases to their customers. Additionally, President Ford proposed to use existing authority to impose a $3 per barrel fee on imported oil and to decontrol the price of domestic oil.

The Democrat-controlled Congress rejected most of the market-oriented provisions of the president's proposal.[21] Decontrol of natural gas prices was rejected, and price controls were reimposed on oil. Mandatory fuel efficiency standards were imposed on the automobile industry. Increased authority for imposing tariffs was not provided, and during negotiations on the final bill, President Ford agreed to lift the $2 per barrel oil import fee that he had already imposed. Although authority was not included for full-scale production from the naval petroleum reserves or development of the military strategic petroleum reserve, the administration did gain the authority it desired to create a domestic strategic petroleum reserve (SPR).

The administration sought enabling legislation for the SPR that would provide wide discretion in the development of an implementation plan. In early 1975 an office had been established within the

Program Integration Division of FEA to push for SPR legislation and begin studies of options for implementation. These studies were only preliminary in nature and would not be completed quickly enough to influence the pending legislation. Consequently, the administration sought flexibility on the four major legislative issues it confronted.

The first was the question of the size of the SPR. Three numbers appeared in the final legislation. The maximum size of one billion barrels, which appeared in the administration's initial proposal, was chosen primarily because it was a round number larger than the largest size the administration expected it would eventually recommend. The minimum size of 150 million barrels, which constituted the Early Storage Reserve (ESR) component of the SPR to be developed within three years of passage, was an estimate made by the FEA of the amount of existing storage capacity that could be secured for the reserve. Finally, there was a presumption in the legislation that the SPR would contain an equivalent of ninety days of petroleum imports (then about 500 million barrels) within seven years. The presumed size reflected the results of a conflict within the administration that was eventually resolved by the president; FEA had argued for a larger size, the Office of Management and Budget (OMB) for a smaller.

The second major issue was whether or not the petroleum industry would be required to hold inventories as part of the SPR. The Senate bill, S.677, initially contained authority for FEA to require importers and refiners to hold petroleum reserves. Although the Interior and Insular Affairs Committee reported S.677 without the industrial petroleum reserve (IPR) provision, Senator Jackson restored it by amendment on the Senate floor after placing in the record a letter from FEA Administrator Frank G. Zarb indicating that the administration did not oppose the inclusion of discretionary IPR authority. Although the administration was opposed to the IPR concept, it was willing to give it further study in return for Senator Jackson's support. The final legislation permitted the FEA administrator to require refiners and importers to store up to 3 percent of their annually refined and imported volumes.

The regional storage of petroleum products was the third major issue. New England senators wanted the SPR to include storage of residual fuel oil in their region. They were successful in amending S.677 to require the creation of a regional petroleum reserve (RPR) to store reserves of imported products that constituted more than 20 percent of regional consumption. The administration was uncertain of the need for the RPR and therefore supported the inclusion of a loophole in the final legislation that gave the FEA administrator discretion to meet RPR requirements with crude oil or petroleum products stored outside the region.

Finally, there was the issue of how to secure oil for the reserve. The administration originally intended revenues from the sale of naval petroleum reserve oil to be a major source of funding for the SPR. Although both the House and Senate passed revisions of HR. 49, which authorized increased production from the naval petroleum reserves, differences could not be reconciled in conference before the end of 1975. Consequently, the SPR legislation was written to give the FEA administrator the authority to use for the SPR any naval petroleum reserve production that was eventually allowed by Congress. Authority to use oil received as royalties for production on federal lands was also granted along with authority to exchange oil and purchase it outright.

The Senate passed S.677 in July by a vote of 91 to 0. On the House side, SPR authority was part of an omnibus energy policy bill, H.R. 7014, passed in September by a vote of 255 to 148. An omnibus energy bill, S.622, emerged from conference almost two months later and was approved by the House on December 15 and the Senate on December 17. Although President Ford objected to a number of its provisions not related to the SPR, he signed it into law on December 22, 1975, as the Energy Policy and Conservation Act (PL94-163, EPCA).[22]

In addition to the provisions already discussed, EPCA required the FEA administrator to submit within ninety days a plan for implementation of the ESR and before December 15, 1976, a plan for implementation of the full SPR program. Each house of Congress retained veto power over the plans and any subsequent amendments to them. EPCA provided an initial standing budget authority of $1.1 billion to begin the SPR program.

A Perspective for Viewing SPR Implementation

Passage of EPCA is only the beginning of the story. Fortunate we would be if words on paper could instantly put oil into storage. Implementation of the SPR program has been plagued with technical, organizational, and political problems that have resulted in delays and cost overruns. One objective of the next part of this volume is to provide an account of the successes and failures of the SPR program. A simple conceptual framework will provide a useful perspective for viewing the SPR experience and relating it to the generic topic of program implementation.

In recent years students of the policy sciences have devoted considerable attention to the problems of program implementation.[23] Unfortunately, the complex, varied, and open-ended nature of the implementation process has not been conducive to the development of general theories with nonobvious implications. Nevertheless, descriptive models can be valuable tools for guiding analysis. For example, the "assembly" metaphor developed by Eugene Bardach focuses attention on the variety of program elements that must be secured and integrated on a timely basis for successful implementation.[24] The descriptive model we will employ views implementation as an evolutionary process during which a succession of different analytical and managerial capabilities must be developed. Four stages in the evolutionary process can be identified: strategic planning, systems design, management control, and project execution.[25]

Strategic planning refers to the translation of a general goal into a specific plan for its execution. For example, the general goal of the SPR program is the stockpiling of petroleum to reduce U.S. vulnerability to disruptions of the world oil market. The plan for its implementation must answer a number of interrelated questions: How large should the stockpile be? What role should the private sector plan in its development? What types of crude oil and petroleum products should be stored? What types of storage facilities should be used? What rates of fill and drawdown should these facilities be able to sustain? What criteria should be used for selecting facility sites? Effective strategic planning requires the skills of generally trained policy analysts who are capable of critically consuming technical information. Although it is most important at the time the program is first initiated, strategic planning usually must continue throughout the implementation process as circumstances change and better information becomes available through experience.

During the *systems design* stage, detailed planning for the program elements must be completed. If stocks are to be held by the private sector, regulatory or subsidy approaches must be developed. If stocks are to be held by the government, sites must be selected, engineering specifications drawn, facilities integrated into a coherent design, and development schedules established. Successful systems design requires personnel with specific technical skills. Experienced engineers are particularly important if the government is going to develop its own facilities.

The development of a *management control* infrastructure is necessary to ensure that the systems design is being carried out in an efficient manner. Procedures must be developed for awarding and monitoring contracts, hiring and training program personnel, scheduling and coordinating construction activities, and budgeting financial resources. Computerized information and reporting systems may have to be developed to assist in these tasks. Capabilities may have to be developed to answer requests for information originating from Congress, the public, or other executive branch agencies. Development of an effective management control infrastructure requires personnel with general managerial experience and advanced technical skills.

The final implementation stage is *project execution*. During this stage the program plan is actually carried out. If the plan calls for the government to build and operate facilities, sites must be obtained through negotiation or eminent domain, environmental permits must be secured, materials must be ordered, and contracts with private firms must be awarded and monitored. Once the facilities are completed, they must be maintained and operated. In the case of stockpiling, the logistics of oil acquisition and drawdown must be managed. Efficient project execution requires middle-and lower-level managers with substantive experience in the tasks being performed.

Problems are bound to arise when attempts are made to accelerate implementation by attempting to begin the stages in parallel rather than sequence. Higher costs are likely to be incurred because activities in the later stages probably will not fully conform to the specifications for overall efficiency set in the earlier stages. For example, ordering equipment with long lead times before the systems design is completed may permit faster construction but may also commit the program to a less efficient design than would have eventually been selected. For many programs it is worth trading higher costs for earlier completion. This trade-off is not always clear, however. Prematurely beginning the later stages may also contribute to unexpected future delays by producing decisions that cannot be reconciled with the specifications that eventually emerge from the systems design stage. Many of the problems encountered by the SPR program can be traced to the early attempt to begin project execution prior to the completion of systems design work and development of an adequate management control infrastructure.

Two important characteristics of public agencies make it difficult to move smoothly from one implementation stage to the next, particularly for new programs. First, it is difficult to maintain staffs that are appropriately skilled. Government salaries may not be competitive with private industry for personnel with highly specialized skills. The civil service system may make it difficult to replace personnel whose skills were important in the early stages with personnel having the skills needed in the later stages. Agency expertise may not even be sufficient to make possible effective use of outside consultants.[26]

Second, the bureaucratic environment of the program may impede the completion of the implementation stages. It may be difficult to secure, on a timely basis, approval for important decisions from agency superiors who are concerned with a large number of other programs as well. Similarly, obtaining concurrences from other agencies may require time-consuming negotiations and bureaucratic maneuvering. Incompletely resolved issues at one stage of the implementation process can cause problems in the next.

The SPR program was hindered by these organizational problems, particularly after the creation of the Department of Energy. Its organizational problems were exacerbated by the failure to develop an effective management control infrastructure before beginning project execution. The resulting problems gave the program a reputation for poor management that placed it at a disadvantage. The reputation for poor management attracted the attention of congressional and executive branch investigators. Responding to the investigations further taxed the already strained managerial resources of the program.[27]

How did the SPR program face these challenges and what was it able to accomplish?

Notes

1. For example, consider the fateful year of 1941. The U.S. embargo of oil to Japan and the threat U.S. bases in the Philippines posed to Japanese supplies of oil from the East Indies were important factors in their decision to

attack the U.S. The decision by the Nazis to invade the Soviet Union was influenced by their desire to directly control the oil fields of Caucasia. The military capabilities of Japan and Germany were eventually substantially reduced by their limited access to oil supplies. Of more relevance to our present situation, some Americans advocated a military response to the 1973 Arab oil embargo, and there has been widespread support for a greater U.S. military role in the Persian Gulf region.

2. U.S. Department of Energy, Energy Information Administration, *Monthly Energy Review*, April 1981, p. 32.

3. For a general discussion, see William D. Nordhaus, "The Energy Crisis and Macroeconomic Policy," *Energy Journal* 1, no. 1 (January 1980): 11–20; for numerical estimates of the magnitude of these effects, see William W. Hogan, "Import Management and Oil Emergencies," in *Energy and Security*, ed. David A. Deese and Joseph S. Nye (Cambridge, Mass.: Ballinger Publishing Company, 1981), pp. 261–301. Hogan estimates that the sum of the balance of payments, inflation, and monopsony components of the import premium may be as low as $1 and as high as $28 per barrel (p. 275).

4. For a well-balanced discussion of appropriate U.S. energy policy, see Hans H. Landsberg, ed., *Energy: The Next Twenty Years* (Cambridge, Mass.: Ballinger Publishing Company, 1979).

5. For a discussion of its history and administrative politics, see Glenn H. Snyder, *Stockpiling Strategic Materials: Politics and National Defense* (San Francisco: Chandler Publishing Company, 1966). Also see Michael W. Klass, James C. Burrows, and Steven D. Beggs, *International Minerals Cartels and Embargoes* (New York: Praeger, 1980).

6. The Aerospace Corporation, "An Overview of the Strategic and Critical Materials Stockpiling Program and Its Relationship to SPR Sizing," prepared for the Strategic Petroleum Reserve Office, U.S. Department of Energy, September 1980, p. A-8.

7. For an historical summary, see Comptroller General of the United States, "Capability of the Naval Petroleum Oil Shale Reserves to Meet Emergency Oil Needs," GAO, October 5, 1972.

8. An offsetting advantage of in situ stockpiling is that it permits drawdowns for extended periods and may thus provide greater deterrence against embargoes than stockpiles with equivalent surge capacity but shorter endurance. See Douglas R. Bohi and Milton Russell, *U.S. Energy Policy: Alternatives for Security* (Baltimore: Johns Hopkins University Press, 1975), pp. 95–99.

9. For a history of this period see William J. Barber, "The Eisenhower Energy Policy: Reluctant Intervention," in *Energy Policy in Perspective*, ed. Craufurd D. Goodwin (Washington, D.C.: The Brookings Institution, 1981), pp. 205–86.

10. Walter J. Mead and Philip E. Sorensen, "A National Defense Petroleum Reserve Alternative to Oil Import Quotas," *Land Economics* 47, no. 3 (August 1971):211-24.

11. David H. Davis, *Energy Politics* (New York: St. Martin's Press, 1978), p. 93.

12. National Petroleum Council, "Emergency Preparedness for Interruption of Petroleum Imports into the United States," Proposed Final Interim Report, July 24, 1973.

13. U.S. Congress, Senate, "Strategic Petroleum Reserves," Hearings before the Committee on Interior and Insular Affairs, 93d Cong., 1st sess., May 30 and July 26, 1973, p. 2.

14. Federal Energy Administration, Office of International Energy Affairs, "U.S. Oil Companies and the Arab Oil Embargo: The International Allocation of Constricted Supplies," prepared for the Subcommittee on Multinational Corporations of the Committee on Foreign Relations, U.S. Senate, January 27, 1975 (Washington, D.C.: Government Printing Office, 1975), pp. 7-8.

15. *Petroleum Intelligence Weekly*, October 20, 1980. p. 11.

16. National Petroleum Council's Committee on Emergency Preparedness, *Emergency Preparedness for Interruption of Petroleum Imports into the United States* (Washington, D.C.: National Petroleum Council, September 1974), pp. 40-42.

17. For a review of the economic impacts of the Arab embargo, see Robert S. Dohner, "Energy Prices, Economic Activity and Inflation: A Survey of the Issues," presented at the Conference on Energy Prices, Inflation and Economic Activity, MIT Center for Energy Policy Research, November 7-9, 1979.

18. Federal Energy Administration, *Project Independence Report*, November 1974, p. 19.

19. Joel Havemann and James G. Phillips, "Energy Report/Independence Blueprint Weighs Various Options," *National Journal Reports* 6, no. 44 (November 2, 1974):1635-54.

20. Thomas H. Tietenberg, *Energy Planning and Policy: The Political Economy of Project Independence* (Lexington, Mass.: Lexington Books, 1976), pp. 91-92.

21. For a general discussion of legislative action on the Ford proposals, see Neil De Marchi, "The Ford Administration: Energy as a Political Good," in *Energy Policy in Perspective*, ed. Craufurd D. Goodwin (Washington, D.C.: The Brookings Institution, 1981), pp. 475-545.

22. The SPR authorities are found in Title I, Part B, Sections 151-66, of EPCA (see Appendix).

23. The work of Jeffrey Pressman and Aaron Wildavsky sparked a revival of academic interest in the study of implementation. See Jeffrey L. Pressman

and Aaron Wildavsky, *Implementation* (Berkeley: University of California Press, 1975). Continuing interest is reflected in recent special issues on implementation in two of the leading policy journals: *Policy Analysis* 1, no. 3 (Summer 1975); and *Public Policy* 27, no. 2 (Spring 1978).

24. Eugene Bardach, *The Implementation Game: What Happens After a Bill Becomes a Law* (Cambridge, Mass.: MIT Press, 1977).

25. Classifications of this sort can be traced to Robert N. Anthony, *Planning and Control Systems: A Framework for Analysis* (Cambridge, Mass.: Harvard University Press, 1965).

26. See, for example, Gary D. Brewer, *Politicians, Bureaucrats, and the Consultant* (New York: Basic Books, 1973).

27. The SPR experience can be contrasted to that of the Navy's Fleet Ballistic Missile Program. The management systems developed for the latter were more successful as shields against interference by other governmental actors than as tools for internal management. Harvey Sapolsky has called this phenomenon the "myth of managerial effectiveness." See Harvey M. Sapolsky, *The Polaris Systems Development* (Cambridge, Mass.: Harvard University Press, 1972), pp. 95-130.

2

Getting Started Under FEA: Strategic Planning

The Energy Policy and Conservation Act (EPCA) provided only tentative goals and general guidelines for implementation of the strategic petroleum reserve (SPR) program. EPCA established the SPR Office within the Federal Energy Administration (FEA) and gave it responsibility for submitting a detailed implementation plan to Congress within one year. The primary effort of the SPR Office during 1976 was to complete the strategic planning that underlies the SPR plan. Cost and schedule estimates aside, the SPR program continues to resemble closely the program proposed in the plan. Consequently, a review of the major strategic planning issues confronted during preparation of the plan provides a sketch of the current SPR program as well as a starting point for understanding its implementation.

Overview of the Strategic Planning Process

Strategic planning for the SPR actually began before EPCA was passed. In early 1975 a small group was assembled within the Program Integration Branch of FEA to work in support of the strategic and naval petroleum reserves legislation being proposed by the administration. The group was headed by Robert L. Davies, who later became the first director of the SPR Office. As a systems analyst in the Department of Defense, he had recommended oil stockpiling as an alternative to requests made by the navy for more ships to protect the growing world supertanker traffic. During the Arab embargo he joined the Federal Energy Office, the predecessor of the FEA, to participate in Project Independence. He was thus familiar with the major strategic planning issues that would have to be addressed.

Although his staff was too small to begin in-house studies, Davies had a budget of about $1 million for contractor support. By the latter part of 1975, general planning and technical studies were underway. One of the technical studies focused on the costs, benefits, and potential sites for salt dome cavern storage. Two other studies looked at above-ground steel tank storage and storage in conventionally mined caverns. The remaining technical study began collecting information that would be needed in preparation of the programmatic environmental impact statement for the SPR. Results from the technical studies were used by the general support contractor to compare storage alternatives. While these studies were underway, the start of the new fiscal year permitted an expansion of the staff from four to about forty, and passage of EPCA resulted in the staff being designated the SPR Office.

EPCA required the SPR Office to submit a plan for implementation of the Early Storage Reserve (ESR) to Congress within ninety days of its passage. Although the studies begun the previous July were not yet completed, a number of general conclusions were reached by the April 22, 1976, deadline.[1] First, the goal of having 150 million barrels of petroleum in storage by the end of 1978 could be achieved only through the conversion of existing salt dome caverns and conventional mines. It was estimated that 200 million to 300 million barrels of existing capacity could be secured and converted to petroleum storage. Second, the goal of having 10 percent of the presumed 500 million barrel SPR in storage within eighteen months was impractical. The goal could only be reached through the leasing of existing steel tank facilities or oil tankers. The former was judged uneconomical, the latter environmentally unsound. Third, the optimal storage modes in terms of flexibility, cost, and speed appeared to be salt dome caverns on the Gulf Coast and converted mines in the South and Midwest. Fourth, site design considerations would make it practical to store only between four and ten distinct grades of crude oil.[2]

Although the ESR Plan did not identify specific sites, the timely completion of site-specific environmental impact statements required that the primary candidates be selected early. The technical studies being conducted by the contractors were a major source of candidate sites. The FEA sought information about prospective sites from the private sector by publishing a notification of its search in the *Federal Register*. Additionally, the FEA had its regional offices along the East Coast conduct informal surveys of existing facilities that might be obtained and had the General Services Administration determine the availability of surplus government-owned facilities. A contractor was hired to prepare site-specific environmental impact statements for eight candidate ESR sites; later the list was expanded to include sites suitable for the development of new caverns.

Looking ahead, the SPR Office anticipated that long lead times for the procurement of equipment and materials, such as pumps and piping, could lead to delays in implementation of the ESR. A decision was made to develop preliminary specifications for critical equipment and materials so early procurement could be initiated prior to completion of site selection and systems designs. Actual procurement began in late 1976, even prior to completion of the SPR Plan. Project execution thus began before strategic planning was completed and before systems design and the development of a management control infrastructure had started.

In the spring of 1976 additional contracts were let for studies that would be needed for completion of the SPR Plan by the December deadline. The Maritime Administration conducted analyses of tanker storage and transportation issues. Other contractors looked at the need for regional reserves, the feasibility of an industrial reserve, local and national economic impacts of the SPR program, and the desirable mix of crude oil types to be placed in storage.

In August 1976 the SPR Office enjoyed a strengthening of its position within FEA. Until then the director of the SPR Office reported to a deputy assistant administrator who in turn reported to an assistant administrator who finally reported to Frank G. Zarb, the FEA administrator. Zarb recognized that a shorter chain of command would be needed for expeditious decision making as the implementation process progressed. He set up the SPR Office as a separate entity within FEA to be headed by an assistant administrator and appointed Thomas E. Noel, a member of his staff, to the job. (Bob Davies stayed on as Noel's chief deputy.) Zarb also personally chaired a committee within FEA to meet on short notice to resolve important strategic planning issues as they arose. The new organizational arrangements not only facilitated completion of the SPR Plan by the EPCA imposed deadline, but also made it easier to secure additional personnel and resources.

Major Strategic Planning Issues

The strategic planning issues that had to be addressed during preparation of the SPR Plan were both numerous and interrelated. Resolution of some key issues greatly narrowed the range of feasible options for resolving others. Some decisions were effectively irreversible; others could and would be reversed later. Technical analysis played an important role but not to the total exclusion of political considerations. Because of uncertainties about technical factors and the nature of future disruptions of the world oil market, many decisions reflect an attempt to retain flexibility for the program.

Ten issues can be identified as most important:

1. Size and Schedule

Although FEA had authority under EPCA to recommend a reserve size of between 150 million and one billion barrels, the legislation carried a strong presumption that a reserve corresponding to about ninety days of petroleum imports would be chosen. Accepting this presumption, FEA selected a reserve size of 500 million barrels. Except for the eighteen-month goal already abandoned in the ESR Plan, the schedule implied in EPCA was affirmed in the SPR Plan: 150 million barrels in storage by December 1978, 325 million barrels by December 1980, and 500 million barrels by December 1982. FEA estimated the budgetary costs of the 500 million barrel reserve to be between $7.5 billion and $8 billion, approximately 90 percent of which would be for the acquisition and initial transportation of oil.

In selecting the 500 million barrel size, the FEA leadership was adhering to the compromise reached with the Office of Management and Budget (OMB) the previous year during preparation of the administration's legislative proposals. Nevertheless, the size issue was investigated analytically by the SPR Office. Cost-benefit analysis suggested that the 500 million barrel reserve could be justified on economic grounds with relatively optimistic assumptions about the severity, likelihood, and duration of oil market disruptions. Even though more pessimistic but not unreasonable assumptions would support larger reserve sizes, the analytical staff realized that the administrator was unwilling to reopen the issue with the president and therefore concentrated on supporting the 500 million barrel goal.

The SPR Plan indicated, however, that a decision could be made at a later date to expand the reserve. Indeed, within three months of its submission to Congress, President Carter announced that his new administration intended to increase the size of the reserve to one billion barrels.

Chapter 6 investigates the size issue in greater depth. In particular, it looks in detail at analyses of SPR size, especially in the context of the bureaucratic battles that were fought within the Carter administration over implementation of the expanded reserve.

2. Use of IPR Authority

The Ford administration was philosophically against establishing a regulatory program to require the petroleum industry to hold strategic reserves for the government. It had accepted discretionary authority to establish an industrial petroleum reserve (IPR) out of political expediency. Within the SPR Office, some analysts initially thought that the implementation of the IPR might provide an avenue for accelerating the SPR schedule. After the completion of a feasibility study by a contractor and the sampling of industry attitudes at a public hearing, however, the SPR Office decided to recommend rejection of the IPR authorities. The difficulties of designing an equitable regulatory system and the likelihood that the IPR would be challenged in the courts as an unconstitutional taking of property without just compensation cast doubt on the proposition that the IPR could actually be implemented more quickly than a program involving government-owned storage. Additionally, the IPR would most likely rely on decentralized storage that would cost more, involve higher environmental risks, and perhaps provide less drawdown flexibility than a centralized storage system.

Chapter 7 reviews the IPR decision in greater depth and considers subsequent attempts to expand the role of the private sector in the SPR program.

3. General Location

The SPR Office concluded that the Gulf Coast is the best general location for storage facilities. The Gulf Coast, itself a major refining region, is linked to the interior of the U.S. by crude oil pipelines. Its terminal facilities for tankers provide an ocean link to refineries along the U.S. coasts and in the Caribbean. Access to this highly developed petroleum transportation infrastructure would provide great flexibility in the acquisition and distribution of SPR oil. Fortunately, these transportational advantages coincided with the availability of existing salt dome caverns and abundant potential for developing new ones.

4. Storage Facilities

Analysis by the SPR Office confirmed the preliminary conclusion expressed in the ESR Plan that conversion of existing solution-mined salt dome caverns and existing conventionally mined caverns to crude oil storage facilities would be the best approach for implementing the first phase of the SPR. Creation of new solution-mined salt dome caverns was identified as the best way to provide storage capacity beyond that provided by suitable existing caverns. In addition to cost, five other criteria were considered in comparing alternative storage modes: technical feasibility, availability of adequate storage capacity on a timely basis, proximity to existing petroleum distribution systems, environmental impact, and the security of stored oil. The major alternatives were new conventionally mined caverns, surface tanks, and tankers. Storage in depleted oil wells, floating rubber bags, and shut-in oil production capacity were also considered, but not seriously, because of technical problems or excessive cost.

Solution mining, which involves leaching salt with water, is a well understood and widely used process. In the U.S. private firms have solution-mined over 900 caverns to provide brine as a petrochemical feedstock and to provide storage capacity for natural gas, propane, and other hydrocarbon products. At the time the SPR Plan was being written, salt dome caverns were already being used for storing crude oil in Europe. Although solution mining is relatively slow (seven barrels of fresh water must be injected and seven barrels of brine removed for each barrel of storage created), existing caverns can be partially filled with oil while they are being expanded through leach-

Table 1
Summary Analysis of Storage Alternatives in the SPR Plan
(December 1976)

	SOLUTION-MINED CAVERNS	*CONVENTIONALLY MINED CAVERNS*	*SURFACE TANKS*	*TANKERS*	*LAGOONS/RUBBER BAGS*	*DEPLETED OIL WELLS*	*DEPLETED SHUT-IN OIL*
Technical Feasibility and Suitability for Storage	Yes	Yes	Yes	Uncertain	Uncertain	No	Yes
Adequate Storage Capacity Available on a Timely Basis	Yes	Yes[1]	Partially	Uncertain	Uncertain	No	Partially
Proximity to Existing Petroleum Distribution System	Yes	Yes[1]	Yes[1]	Yes[1]	Uncertain	No	Partially
Environmental Impact	Low	Low	High	High	High	Low	Low
Security	Good	Good	Poor	Very Poor	Poor	Good	Good
Cost per Barrel							
Existing	$1.10-1.75	$ .90-1.50	$8.00-12.00	Over $6.00	Over $15[2]	Over $45[4]	Over $100
New	$1.35-2.15	$6.00-9.00			Uncertain[3]		

SOURCE: Strategic Petroleum Reserve Office, "Strategic Petroleum Reserve Plan," December 15, 1976. Table IV-1, p. 75.

[1] For at least a portion of the requirements.

[2] Cost per barrel for Rubber Bag storage used by the military for fuel was approximately $15.00 (FY 1968 data).

[3] There is no firm basis for estimating costs of large-scale lined lagoon storage.

[4] This assumes that only about 25 percent of the oil could be recovered in six months. On that basis, the cost would be about equal to the cost of the oil stored and not recovered, or about $45 per barrel.

ing. The direct cost of solution mining was estimated to be between \$1.35 to \$2.15 per barrel for new caverns as compared to \$6.00 to \$9.00 for new conventionally mined caverns—the lowest cost alternative for creating new capacity. Except for brine disposal, the environmental impact of solution mining is minimal. The caverns provide relatively secure storage.

Conversion of existing mines appeared superior to solution mining in terms of direct facility costs and environmental impacts. However, mines tend not to be located near the existing petroleum distribution system and, unlike solution-mined caverns, cannot be filled with oil until the conversion process is completed. Conventional mining of new caverns was rejected as too expensive and slow.

Although the petroleum industry relies primarily on steel tanks for crude oil storage, the SPR Office concluded that little excess capacity would be available for leasing. Once land was acquired for sites, steel tanks could be built relatively quickly. However, the direct costs of development would be much higher, the environmental impacts more severe, and security lower than for underground caverns. The comparison with solution-mined caverns was expected to be less favorable the larger the number of steel tanks constructed because less attractive sites would have to be used.

The direct cost of converting tankers to storage facilities appeared to be lower than constructing steel tanks. The potential environmental impact, however, was judged to be much more severe, including the possibility of hydrocarbon emissions and the danger of spills. There was uncertainty over whether or not tankers could be modified to make them environmentally acceptable for long-term storage. Because of the possibility of collision with other ships and their relatively high vulnerability to sabotage, tankers appeared to be the least secure storage mode. Consequently, the use of tankers was rejected.

5. *Site Selection*

The same general criteria used to compare alternative storage modes were also employed in the site screening process. Investigations of 154 salt dome caverns and 300 mines were initially conducted. Three salt dome caverns with total estimated capacity of 212 million barrels and five mines with total estimated capacity of 181 million barrels were selected as candidate sites for the ESR. An additional thirteen sites potentially suitable for solution mining of new caverns were selected as candidate sites for meeting the balance of SPR storage requirements. Final selection would be made after completion of studies that would provide information needed for estimating the costs of acquiring sites and related rights-of-way, preparing environmental impact statements, and determining the technical suitability of existing caverns for oil storage.

The SPR Office planned to begin acquisition of ESR sites in March 1977. Preliminary negotiations with site and right-of-way owners were already being conducted by the U.S. Army Corps of Engineers on behalf of the FEA. Lease agreements as well as purchases were being considered. If negotiations failed, the corps would provide the necessary documentation to support condemnation proceedings.

6. Selection of Petroleum Stocks

The SPR Office had to decide if petroleum products as well as crude oil would be held in the reserve. Storing crude oil rather than specific petroleum products offered two major advantages. First, because the slate of products obtained from any particular crude oil can be altered during refining, storage of crude oil offered greater flexibility for meeting product shortfalls. Second, the life-cycle costs of storage are greater for products than for crude. Residual fuel oil usually must be heated during drawdown, and stocks of lighter products must be rotated to avoid deterioration. Consequently, a decision was made not to store petroleum products.

The specification of the particular types of crude oil to be stored was a more complex problem. Crude oils differ in terms of the products they yield from simple distillation, their content of sulfur and other impurities, and a number of other physical properties relevant to refining. The wellhead prices of crude oils reflect these characteristics and transportation costs to markets.[3] Analysts also had to be concerned about the availability of crude types during acquisition; mixing crudes with widely different properties was considered undesirable, and increasing the number of segregated storage caverns would result in higher facility costs.[4]

Specifications were established for six different crude types ranging in quality from intermediate density/high sulfur to low density/low sulfur. Linear programming models of the U.S. refining industry were then used to determine what combinations of crude types would provide the best protection against a range of hypothesized disruption scenarios. Assuming restrictions against the burning of high-sulfur fuel oil would not be lifted during disruptions, it appeared that the most flexible and cost-effective oil mix would include about 60 percent intermediate density/ high sulfur crude with the remainder light density/low sulfur. It was anticipated that these crude types would be readily available in the world market until the early 1980s.

7. Oil Procurement

Four options for procuring oil for the SPR were considered: purchase in the world market without entitlement credit, purchase in the world market with entitlement credit, use of royalty oil, and use of naval petroleum reserve oil. The use of royalty oil was rejected for political and technical reasons. Small refiners enjoyed a financial advantage from preferential access to oil collected by the government as royalties for production on public lands; they most certainly would have vocally opposed loss of their special treatment. Because the royalty oil was widely dispersed and of heterogeneous quality, an administratively complex system of trades would have been required to secure delivery of oil appropriate for the SPR. Use of the naval petroleum reserve oil would also have required extensive trading and would have resulted in losses of revenue comparable to expenditures needed to purchase equivalent amounts in the world market.

The SPR Office viewed purchasing in the world market as administratively and budgetarily superior to use of government-owned domestic oil. There was disagreement within FEA, however, over whether or not SPR purchases should be incorporated into the entitlements system, which equalized the average acquisition cost of crude oil to refiners by granting importers a right to purchase price-controlled domestic oil. Incorporating SPR purchases into the entitlements system would raise the national average acquisition cost and hence raise product prices but nevertheless would lower the budgetary (but not the real economic) costs of oil acquisition for the SPR by several dollars per barrel.[5] The decision to modify the entitlements system to give SPR purchases an entitlement credit was influenced by the preferences of OMB for its smaller budget impact.

An agreement was reached with the Department of Defense to have its Defense Fuels Supply Center actually procure oil for the SPR. To avoid delay in the filling of available storage capacity, it was anticipated that solicitation of bids for oil would begin six months prior to opening of the first facility. Because pipeline connections between SPR sites and terminals would be completed after storage capacity started to become available, the SPR Office planned to use higher-cost barge transportation for initial deliveries.

Reluctance of the Carter administration to resume purchase of oil for the SPR in the wake of Iranian production cutbacks led to reconsideration of acquisition alternatives. Chapter 4 reviews the acquisition question and the political controversy surrounding it.

8. Drawdown and Distribution

It is not surprising that the SPR Office devoted minimal consideration to when and how the SPR oil should be used. Analysis of issues more relevant to getting the program started had greater claims on scarce resources. The maximum sustainable drawdown rate for the SPR system had to be determined prior to the design of facilities. Because the marginal cost of incorporating faster drawdown capabilities was estimated to be low, a relatively high maximum drawdown capability of 3.3 million barrels per day was selected. This rate would permit the entire reserve to be drawn down in less than six months.

The SPR Plan was vague about when the SPR drawdown should begin. Although it argued that the establishment of "trigger" conditions was undesirable in light of the large number of factors that should be taken into account in the drawdown decision, it offered little guidance on the weights that these factors should be given. Little was said about how the rate of drawdown should be determined. No indication was given as to how SPR oil would be priced and allocated during drawdowns except for the statement that the decisions would be consistent with the Emergency Petroleum Allocation Act and other regulatory authorities. As discussed in Chapter 8, little progress was made in subsequent years in resolving the drawdown and distribution issues.

9. Regional Storage

The most controversial SPR issue within FEA was whether or not a regional petroleum reserve (RPR) should be established. There is a strong presumption in EPCA that the SPR would include regional storage of refined products for which imports constituted 20 percent or more of regional consumption. Under this criterion, the RPR would consist of storage of residual fuel oil along the East Coast. However, EPCA permits crude oil stored in a nearby region to be substituted for regional storage if it will result in comparable protection for the region. After several rounds of debate, the SPR committee within FEA decided not to implement the RPR.

Analysis conducted by the SPR Office suggested that during disruptions there would be adequate capacity available for refining SPR crude and shipping the resulting residual fuel oil to the East Coast. It therefore was argued, with OMB support, that the more costly regional storage of residual fuel oil was unnecessary. Advocates of the RPR worried about the possibility of severe disruptions during winter months when private inventories might not be large enough to meet basic needs while the SPR crude was being drawn down, refined, and transported. Some anticipated the political consequences of not implementing the RPR. Attempts by New England congressmen to reverse the initial RPR decision are considered in Chapter 8.

10. Implementation Process

The SPR Office planned to rely heavily on contractors during the systems design, management control, and project execution stages of implementation. The SPR Office staff, already numbering 125, would expand to only about 150. During the development of facilities, the staff would administer FEA contracts with an executive engineer, a construction manager, several architect/engineers, and a large number of construction contractors.

The executive engineering firm would have responsibility for coordinating systems design activities. This would include: developing technical criteria, work scopes, and cost estimates for facility designers; testing caverns and facilities; and supporting the procurement of long lead-time equipment. The executive engineers would also work with the Corps of Engineers on acquisition of real estate.

The construction manager would have responsibility for providing technical assistance in the management of construction contracts. This would include: implementation of a construction management and cost control system; on-site inspection and coordination service; evaluation of construction bid packages; and coordination of the delivery and storage of long lead-time equipment. The construction manager would provide many of the services, except subcontracting, normally provided by a prime contractor. FEA would directly contract with all parties, including the firms that would actually do the construction work.

The architect/engineering firms would provide detailed designs for facilities. It was anticipated that one firm would handle all pipeline design work and another all design work for conventionally mined sites. One or more architect/engineers would design facilities for the solution-mined sites.

The relationships of the anticipated implementation actions are shown in Figure 1. At the time the SPR Plan was submitted, the actions leading up to and including preliminary site selection were completed. Negotiations with site owners were underway. Preliminary facility design work had begun, and orders were being placed for

Figure 1
SPR Plan: Major Implementation Actions

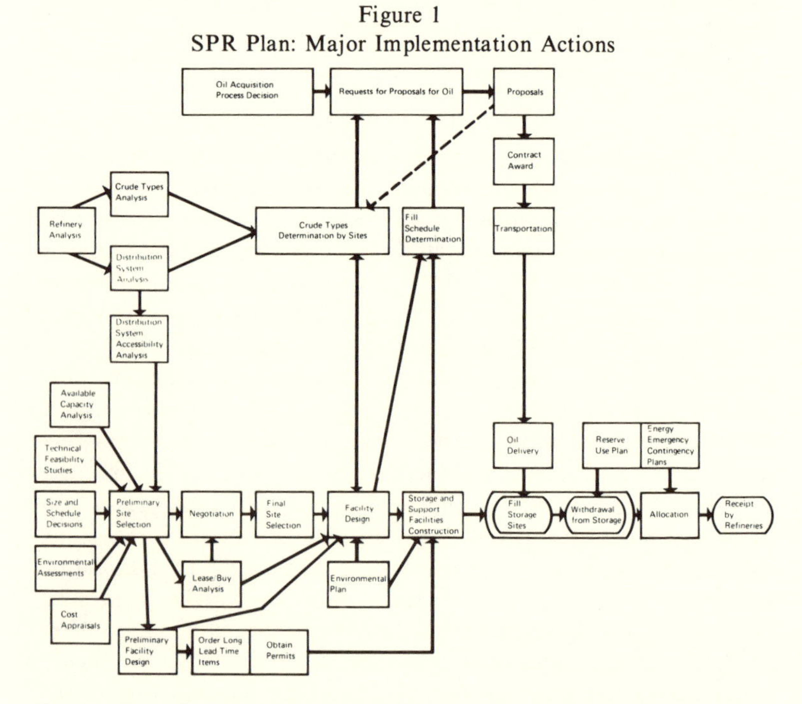

SOURCE: Strategic Petroleum Reserve Office, "Strategic Petroleum Reserve Plan," December 15, 1976. Table VIII-3 p. 171.

long lead-time equipment. It was anticipated that construction contracts would begin to be awarded in the spring of 1977 so that oil fill could commence at some sites as early as August 1977.

Congressional Approval

The SPR Plan was submitted to Congress by the FEA on December 15, 1976.[6] A number of organizations raised objections to the plan in congressional hearings held the following February. The most serious opposition, however, came from senators and congressmen who favored the establishment of regional and noncontiguous reserves.

In a report to Congress, the General Accounting Office (GAO) identified several issues it believed deserved further analysis.[7] One was the extent to which crude oil and petroleum product inventories held by the private sector could be drawn down during emergencies. The analysis of the size issue in the SPR Plan did not explicitly account for the use of private stocks during disruptions. The GAO found this curious in light of the assertion made by the Department of State to the International Energy Agency that the U.S. was meeting its stockpiling obligations under the International Energy Program Agreement (which also established a system for sharing oil during market disruptions and embargoes) through industry reserves. The SPR director argued that the SPR was intended by Congress to provide stocks in excess of those normally held by industry but nevertheless indicated that he was in the process of initiating a major study of the extent to which drawdowns of private inventories could offset major disruptions.

Other GAO comments dealt with the procurement and financing of oil. The GAO urged reconsideration of the use of royalty oil because if offered an opportunity for reducing the budgetary costs of the program. It also suggested financing oil purchases through a tariff on crude oil imports or an excise tax on petroleum products as an alternative to use of general revenues. The SPR director reiterated the reasons for not using royalty oil and noted that studies of alternative financing mechanisms were already under way.

Representatives of the National Wildlife Federation and a Louisiana-based environmental organization expressed concern that the transportation and storage of SPR oil might result in spills damaging to the local ecosystem.[8] They argued that the coastal wetlands of Louisiana were particularly vulnerable to ecological damage from oil spills and therefore were inappropriate locations for SPR facilities. Because the region was already heavily used by the petrochemical industry, the members of the relevant congressional committees discounted these arguments and accepted as adequate the safeguards presented in the environmental impact statements prepared by the FEA.

The Marine Engineers' Beneficial Association argued that storage in tankers, which incidently would provide employment for association members, would be less expensive than storage in salt dome caverns.[9] The specific proposal made was to have all ESR storage, including regional product storage, provided by clusters of tankers that remained seaworthy. The SPR Office did not address the assertion made by the Marine Engineers' Beneficial Association that tanker storage capacity could be made available at a rate much faster than salt dome cavern capacity but presented an Institute of Defense Analyses study that estimated the present value of costs over a fifteen-year program life to be $1.91 per barrel for salt dome caverns and $6.42 per barrel for tanker storage.[10]

Although in general agreement with the SPR Plan, the Petroleum Industry Research Foundation had two specific criticisms.[11] First, the schedule should be accelerated so that 500 million barrels would be in storage by the end of 1980 rather than the end of 1982. Second, the entitlements system should not be used to lower the budgetary cost of SPR oil. The SPR provides national security benefits and therefore should be financed out of general revenues rather than through an implicit tax on consumers. The FEA responded by indicating that the new administration was considering acceleration of the SPR schedule and reconsidering the use of the entitlements system. Additionally, in response to questions raised during his testimony, the director of the Petroleum Industry Research Foundation advocated storage of fuel oil in New England.

The most serious attack against the SPR Plan was led by Senator Edward M. Kennedy of Massachusetts. Drawing on research conducted by the New England Regional Council and the Energy Policy Office of the Commonwealth of Massachusetts, he argued vigorously for storage of residual fuel oil in New England.[12] In testimony before the Interior and Insular Affairs Committee of the Senate, he attacked the FEA conclusion that no RPR was needed on several grounds.[13]

First, because the FEA did not provide a detailed distribution plan, it was not possible to judge the adequacy of the proposed SPR system for providing residual fuel oil to the Northeast. No guarantees were given that SPR oil refined in the Caribbean would actually result in shipments to New England. Second, the FEA used unrealistically low estimates of import levels of residual fuel oil into New England over the next decade. Third, the FEA failed to document that adequate barge and tanker capacity would be available on a timely basis to SPR transportation needs during disruptions. Fourth, the FEA overestimated the cost of storing fuel oil in hard rock mines in New England, particularly if No. 4 fuel oil, which does not require heating during drawdown, was stored instead of residual fuel oil. Senator Kennedy was supported in his protest by many other members of the New England congressional delegation.

Regional concerns also were expressed by the senators from Hawaii. Just as EPCA had a presumption for regional storage of products, it also had a presumption that crude oil would be stored in noncontinguous areas of the U.S., including Hawaii. The SPR Office had concluded that it would be more efficient to protect noncontiguous areas with crude stored in the central SPR facilities. Senator Spark Matsunaga presented a plan prepared by the State Energy Office of Hawaii calling for the location of 10 million barrels of SPR storage in steel tanks in Hawaii.[14] He suggested that the SPR Plan be modified to incorporate noncontiguous storage.

The new FEA administrator, John F. O'Leary, was able to deflect these attacks. He assured critics that the new administration would continue to study the regional and noncontiguous storage questions and submit amendments to the SPR Plan if warranted. However, he urged Congress not to wait until these studies were completed before approving the SPR Plan so as not to delay implementation. His arguments prevailed. Senator Kennedy did not push his resolution of disapproval and the House likewise acquiesced. Consequently, the SPR Plan became effective on April 18, 1977.

Notes

1. Strategic Petroleum Reserve Office, "Early Storage Reserve Plan," April 22, 1976.

2. Although crude oils have a large number of physical characteristics relevant to refining, sulfur content and specific gravity are the most promi-

nent. Sweeter (lower sulfur content) and lighter (lower specific gravity) crudes generally command higher market prices, other things being equal, because they have greater natural yields of the more valuable distillation products.

3. Low density/low sulfur crudes command a premium because they have larger natural yields of higher valued light, clean products. Oil from fields with lower transportation costs to market also command a premium at the wellhead.

4. Crude oils can be blended. It was feared, however, that mixing in the cavern would be incomplete so that the characteristics of the oil being withdrawn would fluctuate. This would make it difficult for refiners to swap oil so that it could be refined in the most efficient facilities. In order to maintain high drawdown rates, it was determined that no more than two crude types be stored at any single facility.

5. The entitlement benefit was then about $2.25 per barrel.

6. Strategic Petroleum Reserve Office, "Strategic Petroleum Reserve Plan," December 15, 1976. Its official receipt was not until February 16, 1977, because the original submission inadvertently was not assigned an energy action number as required.

7. Comptroller General of the United States, "Issues Needing Attention In Developing the Strategic Petroleum Reserve," GAO, EMD-77-20, February 16, 1977.

8. U.S. Congress, House, "Strategic Petroleum Reserve," Hearing before the Subcommittee on Energy and Power of the Committee on Interstate and Foreign Commerce, 95th Cong., 1st sess., February 16, 1977.

9. U.S. Congress, Senate, "Review of the Strategic Reserve Plan," Hearing before the Committee on Interior and Insular Affairs, 95th Cong., 1st sess., February 4, 1977.

10. Norman J. Asher and Wendy West, "Use of Tankers for Stockpiling Petroleum," Institute for Defense Analyses, Paper P-1241, January 1977, p. S-1.

11. U.S. Congress, Senate, "Review of the Strategic Reserve Plan," pp. 103-10.

12. The New England Regional Council included the New England governors, representatives of regional commissions, and the FEA's own regional director.

13. U.S. Congress, Senate, "Review of the Strategic Petroleum Reserve Plan," pp. 2-14, 131-230.

14. Ibid., pp. 14-19, 37-103.

3

Debacle Under DOE: Sources of Implementation Failure

Implementation of the congressionally approved SPR Plan began at a fast pace in April 1977. Within a week of the date the plan became effective, condemnation proceedings were completed to acquire three solution-mined storage sites. In July injection of oil into caverns at one of the sites began. By the end of the year, oil was being put into storage at the three solution-mined sites, and a conventionally mined site had been acquired. Meanwhile the new administration announced its intentions to accelerate and expand the SPR program. Both in terms of implementation and policy, the SPR program appeared to be off to a strong start.

During 1978, however, it became apparent that the SPR program could not meet its schedule. First within the SPR Office itself and eventually among the leadership of the newly created Department of Energy (DOE) and the relevant congressional oversight committees, the conclusion was reached that the accelerated schedule was totally unrealistic for the existing program structure. In fact, it became clear that the original schedule and cost estimates were overly optimistic. In November 1978 the SPR Office established a more realistic schedule calling for the availability of 500 million barrels of storage capacity by the end of 1985 rather than by the end of 1982 as was required to meet the goal set out in EPCA and the original SPR Plan.

What went wrong? What are the implications of the early failures for the continuing SPR program?

Sources of Early SPR Program Failure

The problems encountered during implementation of the SPR program spring from a number of sources that are technical, organizational, and political in nature.[1] Although the sources may appear unrelated and peculiar to the SPR, their importance stems from a common feature of public organizations that is generally relevant to program implementation. Namely, managerial and technical expertise resources are scarce and inflexible. Expertise is often limited in both private and public organizations. But inflexibility tends to be greater in public organizations for several reasons: authorized personnel ceilings generally change only with the fiscal year; it is difficult to fire and usually impossible to lay off employees; and the civil service system limits the rewards and sanctions that can be used to attract and motivate skilled employees. When budgets permit, public organizations may attempt to secure needed expertise from consultants on a contractual basis. Securing consultants and monitoring their performance, however, places added demands on managerial resources. Inflexibility is likely to be a particularly critical problem for public organizations implementing new programs on an expedited schedule; it impedes smooth transition from strategic planning to systems design, the development of management infrastructure, and project execution.

Each of the following seven major sources of implementation failure for the SPR program are related in some degree to the central problem of the scarcity and inflexibility of expertise resources in public organizations.

1. Gambling on a Cost/Schedule Tradeoff

Although in recent years the SPR Office has been oriented toward providing low-cost storage at a rate consistent with engineering efficiency, this was not the case in 1977 and early 1978. In order to gain the benefits of stockpiling sooner and to avoid costs from higher oil prices expected in the future, the SPR Office placed a high priority on getting oil in the ground quickly. In doing so, the SPR Office anticipated that somewhat higher costs would result. Implicitly, it was gambling that the interim measures taken to achieve early oil fill would not adversely affect the long-term schedule. Higher costs did result, and to some extent the gamble was lost.

Interim systems were developed for putting oil into storage at the three sites acquired in April 1977. One of the sites was connected to its terminal by a short pipeline that was completed by the end of the year. At the other two sites, temporary pipelines were laid to nearby waterways serving barge traffic. Oil was moved from tanker terminals to these sites by barge until direct pipeline links were completed in late 1978. The barging allowed, albeit at a slow rate and high cost, oil fill to begin at the two sites almost a year earlier than otherwise would have been the case. Unfortunately, it probably contributed to overall schedule slippage by diverting staff resources from design and construction of the permanent facilities that would have permitted oil fill to proceed at a much faster rate. Additionally, it now appears that the ad hoc barging, coupled with weak management control, permitted some oil suppliers to defraud the government by substituting as much as 9 million barrels of low-grade oil for the higher grades specified in contracts.

In order to assure a steady availability of oil for fill, procurement was initiated under the assumption that planned fill rates would be achieved. When technical difficulties prevented the rates from being realized, scheduled tanker and barge deliveries had to be postponed. The resulting demurrage charges increased program costs by approximately $7 million by the end of 1978. More conservative delivery scheduling might have greatly reduced these costs but also might have delayed the filling of available capacity.

Attempts to avoid delays through the early procurement of long lead-time equipment were not fully successful. Used and nonstandard pumps were purchased for the interim fill system. They were difficult to maintain and therefore subject to frequent breakdowns. Because many of the pumps for the permanent systems were purchased prior to completion of systems design work, the architect/engineers had to design around them. Lack of standardization later contributed to maintenance problems. In terms of our four-stage model of implementation set out in chapter 1, the SPR Office began project execution during the strategic planning stage. Decisions of a technical nature were made by a staff consisting predominantly of generalists whose understanding of the organization of the petroleum industry and expertise in economics and policy analysis were needed for timely completion of the SPR plan. Personnel with engineering and managerial expertise were gradually added to the core of strategic planners. However, because of the absence of a strong management control infrastructure, much of their time had to be devoted to troubleshooting the numerous problems encountered during project execution. As a result, the development of appropriate management systems was further delayed.

2. The Contracting System

The most serious mistake made during strategic planning for the SPR was the decision to have the FEA directly contract for construction services. In addition to numerous contracts for analytical, design, operations, and management services, the SPR Office was required to initiate and monitor eighty-five major construction contracts during the first year of site development. The Washington-based staff was simply too small and inexperienced to effectively handle the task.

Lack of expertise would not necessarily have been as serious a problem under a less demanding schedule. Program staff could have gained experience from handling a limited number of initial contracts. Effort could have been devoted to developing management information systems and other management aids. Instead, available resources were consumed by the rapid pace of project execution to the neglect of the development of an effective management control infrastructure.

Initially, less support than anticipated was provided by the firm hired to manage construction. The firm had to hire and train personnel. It also had to expand its expertise in the area of solution-mining technology. Operating under fewer constraints than the SPR Office, it was able to deploy trained personnel at a fairly rapid pace. Nevertheless, its need to gear up for the project limited the extent to which the SPR Office could borrow expertise during the first year of implementation. Additionally, it did not have a strong financial incentive to closely monitor costs because the construction contractors dealt directly with the FEA on questions of cost overruns.

Within the FEA a special procurement board was established to deal exclusively with SPR program contracts. On one hand, the board was hampered by a lack of experience with large-scale construction projects that contributed to cost overrun problems. Prior to the SPR program, most FEA procurements were for studies and research rather than construction. On the other hand, the board successfully expedited the procurement process. It was completing many competitive procurements within two months and sole source procurements in a matter of weeks.

In October 1977 the FEA was merged with the Federal Power Commission (FPC) and the Energy Research and Development Administration (ERDA) to form the Department of Energy. Because ERDA had the largest procurement office, it took over procurement responsibilities for the new department. The ERDA system, designed to handle procurements for large-scale and complex research and development projects, was extremely cumbersome. It posed special difficulties for the SPR program, which had to rely on firms in a sector of industry largely unfamiliar with government procurement processes. Procurements that would have taken a month or two under the FEA system took as long as nine months. The SPR Office diverted staff resources to preparing analyses in support of waivers needed to expedite critical contracts. A few such attempts were successful, but there was almost a six-month period during which new procurements virtually ceased. The new procurement system probably reduced waste; it most certainly delayed the completion of construction of permanent facilities.

During the fall of 1977, the acting SPR Office director, Carlyle Hystad, spent much of his time trying to resolve the procurement problem. He first proposed that construction and associated procurement responsibilities be transferred to the Oak Ridge Operations Office of DOE. But this approach had to be dropped when the DOE under secretary appeared to be unwilling to make a decision in the face of opposition from the head of procurement.

As an alternative, Hystad proposed the establishment of a project office in New Orleans that would have responsibility for construction and associated procurement at the already planned facilities. Anticipating the difficulty of staffing a new office in New Orleans, he wished to keep as much design work in Washington as possible. The proposal was reviewed for the under secretary by a committee headed by a management expert from the Office of the Controller. The under secretary accepted the recommendation of the committee that a Project Office be established with all design responsibility. The Program

Office would remain in Washington and continue to have responsibility for strategic planning, budgeting, and environmental matters. At about the same time, a decision was made to place much of the responsibility for construction contracting in the hands of the construction management contractor; instead of being under direct contract to DOE, construction firms would be subcontractors to the construction manager.

The Project Office was created organizationally in March 1978 and moved to New Orleans the following May. By December 1978 the Project Office was expected to have 150 of the total program staff of 206; by the end of 1978 the New Orleans staff had not yet reached 120. As anticipated by Hystad, many staffers left the program rather than move to New Orleans. Temporary assignment of personnel to the Project Office forced a six-month halt in the planning and design of capacity expansion beyond 250 million barrels. Because government salaries are not competitive with industry salaries for personnel with experience in petroleum operations management and because many view the New Orleans climate as inhospitable and Louisiana schools as inadequate, staffing the Project Office has continued to be a problem since its creation. Even maintaining an experienced guard force at the storage sites has been difficult; once trained, many guards move to higher paying jobs in the local petrochemical industry.

Reorganization and the accumulation of experience by the program staff have greatly reduced procurement problems. Some problems remain. Reliance on the General Services Administration to negotiate contracts with power companies for electricity has resulted in some delays. Cost overruns still occur, although less frequently, and the SPR Office is occasionally embarrassed by bad procurement decisions. For example, it recently purchased an oil-skimming boat with a top speed slower than the current in the part of the Mississippi River where it was intended to be used. Nevertheless, these problems are trivial in comparison to the procurement morass encountered during the critical first eighteen months of program implementation.

3. The Acceleration Decision

James R. Schlesinger served as President Carter's chief energy advisor in the early days of the new administration. Schlesinger viewed reliance on Middle Eastern oil as a serious weakness in U.S.

defense posture. He believed that the then current "softness" of the world oil market would be replaced by very "tight" market conditions beginning in the early 1980s. Consequently, he recommended to the president that the SPR program be expanded to an ultimate size of one billion barrels with the first 500 million barrels of oil to be put in storage two years sooner. The president accepted the recommendations.

In April 1977 the SPR Office received word that Schlesinger wanted to accelerate the schedule. Thomas Noel, the FEA assistant administrator for the SPR, proposed a modest acceleration: 180 million barrels rather than 150 million barrels by the end of 1978, and 500 million barrels by the end of 1981 rather than the end of 1982. Schlesinger demanded a more ambitious schedule, calling for 250 million barrels in storage by the end of 1978 and 500 million barrels in storage by the end of 1980. Noel was forced to accept the Schlesinger schedule. His staff had to divert attention from other matters to try to devise a plan for meeting the accelerated goals. In May an amendment to the SPR Plan was transmitted to Congress establishing the Schlesinger schedule.[2] The acceleration would be accomplished by acquiring additional sites with existing caverns, by increasing brine disposal rates through construction of a pipeline to the Gulf of Mexico, and by accelerating construction of a tanker terminal and pipeline connection to the Weeks Island site, replacing the slower barging of oil sooner.

The acceleration decision reflected an unrealistic view of the capabilities of the SPR Office. Perhaps Schlesinger's experience as secretary of defense led him to believe that all large government projects have slack resources that can be redirected to productive use. Such may be generally true for projects being executed by well-established organizations with experienced personnel. But it was not true for the SPR Office in 1977. Personnel resources were already being strained. Expecting both an acceleration and expansion without a substantial increase in personnel and strengthening of the program infrastructure was unreasonable.

The acceleration decision was detrimental to implementation in three ways. First, it diverted effort from the design of permanent facilities. In particular, completion of planning for pipelines was delayed by diversion of resources to preparation of the amendment. Second, it forced the SPR Office to gamble more heavily on the tradeoff between cost and schedule adherence we have already discussed. Third, it contributed to a loss of credibility that later made securing necessary resources for the program more difficult.

4. Planning Inadequacies and Technical Problems

The preliminary design and cost estimates incorporated in the SPR Plan were based on feasibility studies that did not completely describe required facilities or anticipate the range of technical problems actually encountered during implementation. The feasibility studies were completed by program staff with virtually no familiarity with solution-mining technology and little experience with petroleum operations. Although most of the contractors enjoyed greater levels of technical expertise, none had extensive experience with large-scale solution-mining projects. In fact, there was not a great stock of expertise available because most large caverns were developed as the by-product of the production of brine as a chemical feedstock rather than as storage facilities. It is not surprising, therefore, that the feasibility studies had serious deficiencies.

During detailed design and construction, it was necessary to add a number of neglected elements to the preliminary plans, including fire-fighting systems, marine terminal facilities, and containment dikes around cavern entry wells. Many more cavern entry and brine disposal wells would be needed than initially assumed. The planned safety and security, instrumentation, drawdown, and distribution systems were discovered to be inadequate.[3] Drilling costs turned out to be more than double the initial estimates. The SPR Office was thus confronted with having to build more complex and costly facilities than originally planned.

Two unanticipated technical problems increased program costs and contributed greatly to schedule slippage: the failure of several existing caverns to hold pressure and limitations on the rate of brine disposal. The first phase of the program consisted of injecting oil into existing caverns. For each barrel of oil injected, a barrel of brine had to be removed from the cavern and disposed of. The second phase consisted of creating new cavern capacity by injecting fresh water into the cavern and removing brine. For each barrel of new capacity created, approximately seven barrels of fresh water had to be injected and seven barrels of brine removed. Thus both the rate of oil injection and the rate of new capacity creation were governed by the maximum rate of brine disposal. Further, each substitution of a barrel of new storage capacity for a barrel of existing capacity caused more than a seven-fold increase in the amount of brine that had to be disposed of.

As indicated in table 2, useable capacity at each of the Phase I sites turned out to be less than initially estimated. At Bryan Mound and Sulphur Mines the amounts lost were relatively small—approximately two million to three million barrels at each site. At West Hackberry the loss was almost ten million barrels. Fourteen million barrels less than initially estimated became available at Weeks Island, a conventional salt mine, and utilization was significantly delayed to permit Morton Salt Company to continue mining operations. The most severe loss was at Bayou Choctaw where 38 million barrels of capacity could not be used because several caverns failed to hold pressure. An additional 20 million barrels of planned capacity at Bayou Choctaw were not acquired because of opposition from the State of Louisiana to the harm that would have been caused to the local petrochemical industry through loss of storage capacity needed for feedstock and hydrocarbon products. In all, 87 million barrels of anticipated Phase I capacity turned out not to be available.

The feasibility studies conducted in 1976 envisioned two methods of brine disposal. In the long run brine would be disposed of through pipelines built several miles into the Gulf of Mexico. In the short run brine would be disposed of through deep-injection wells that pumped it into porous rock below the freshwater table. The early part of the fill schedule was dependent upon the deep-injection disposal. Its importance to the entire schedule was increased by delays in the completion of facilities for disposal in the Gulf of Mexico caused by problems encountered in obtaining permits from the Environmental Protection Agency. It was originally estimated that each well could dispose of an average of 30 thousand barrels of brine per day. In practice, however, rates only half as great or less could be sustained. Consequently, oil could not be injected into existing caverns at the rates originally planned. As one program director commented after a short time on the job, "Five months ago I did not know what the hell brine was. Now I live with it."[4]

Although unanticipated technical problems continue to be encountered, the SPR Office, through its accumulation of experience, is better able to deal with them than it was during the first two years of the program. For example, when leaching for Phase II

Table 2
SPR Phase I Facilities (248 million barrels of capacity)

FACILITY	*ACQUISITION DATE*	*INITIAL FILL DATE*	*CONSTRUCTION COMPLETION DATE*	*PIPELINE CONNECTIONS*	*ESTIMATE OF CAPACITY AT ACQUISITION*	*ACTUAL AVAILABLE CAPACITY*	*NOMINAL CONSTRUCTION COST PER BARREL OF CAPACITY*	*PHASE II EXPANSIONS (NEWLY LEACHED CAVERNS)*
Bryan Mound Brazoria County, TX (solution mined caverns)	April 1977	October 1977	June 1979	Seaway Docks and Pipeline	63 mmb[1]	60 mmb	$1.94	120 mmb
West Hackberry Cameron Parish, LA (solution mined caverns)	April 1977	July 1977	June 1979	Sunoco Terminal & Texoma Pipeline	61 mmb	51 mmb	$3.37	160 mmb
Bayou Choctaw Ibernille Parish, LA (solution mined caverns)	April 1977	December 1977	December 1979	St. James Terminal & Capline Pipeline	94 mmb	36 mmb	$3.78	10 mmb

Weeks Island Iberia Parish, LA (conventionally mined caverns)	September 1977	November 1980	September 1979	St. James Terminal & Capline Pipeline	89 mmb	75 mmb	$2.61	0
Sulphur Mines Calcasieu Parish, LA (solution mined caverns)	February 1979	August 1981	December 1980	West Hackberry	24 mmb	22 mmb	$3.07	0
St. James Terminal St. James Parish, LA (terminal, docks, steel tanks)	May 1978 and July 1978	N.A.[2]	December 1979	Capline Pipeline	N.A.	4 mmb (includes all pipeline fill)	N.A.	N.A.

SOURCES: Strategic Petroleum Reserve Office, "Annual Strategic Petroleum Reserve Report," DOE/US-0003, February 16, 1979, pp. 18-19, 42-45; Strategic Petroleum Reserve Office, "Strategic Petreleum Reserve Plan," December 15, 1976, pp. 80-84; and interviews with Strategic Petroleum Reserve Office staff.

[1] mmb-million barrels

[2] N.A.-not applicable

capacity finally began in March 1980 at Byran Mound, it was discovered that the salt contained a higher level of relatively insoluble minerals than anticipated. The minerals tended to plug wells and otherwise slow the leaching process. The SPR Office has been able to direct well workovers and arrange disposal of the resulting material without substantial delay.

5. *Local Politics and Environmental Permits*

The SPR Office has had to devote substantial resources to meeting environmental regulations and obtaining environmental permits. By the end of 1978, the SPR Office completed a programmatic environmental impact statement (EIS), nineteen site-specific EISs, and four EIS supplements in compliance with the National Environmental Policy Act of 1969 (PL91-190). During the same period, four major permits were obtained from the Environmental Protection Agency (EPA) for discharge of substances into surface waters; fourteen permits were obtained from the Corps of Engineers for dredge and fill operations in navigable waters and wetlands; eight permits were obtained from state regulatory agencies; and numerous additional permits were pending. Under the best of circumstances, satisfying all the necessary environmental requirements on a timely basis for the ambitious SPR schedule would have been difficult. For the SPR Office with overextended staff resources and facing local political opposition, it was impossible, especially in the absence of active support from the DOE leadership in resolving the permitting problems.[5]

The major source of local opposition was the extensive use of eminent domain powers by the federal government. All the Phase I sites were obtained through condemnation. Because the Corps of Engineers made appraisals lower than the amounts commonly paid by the pipeline industry, the majority of easements for the 150 miles of pipeline right-of-way obtained during 1978 were done so through eminent domain proceedings. In addition to engendering local hostility, the proceedings themselves contributed to increased program costs and unexpected delay.

The local hostility created a favorable political environment for the governor of Louisiana, Edwin W. Edwards, to seek concessions from DOE in return for cooperation with the SPR program. The permitting process followed by the Corps of Engineers gave Governor

Edwards great leverage in dealing with DOE. If a governor objects to a permit for a project in his or her state, corps regulations require that the permit request be referred by the district office through the division office to the chief of engineers for final determination. The objections need not be supported by evidence and are not strictly limited to environmental concerns. The entire review process can extend over several years if objections are raised.

In December 1977 after all environmental concerns raised by state and local agencies had been resolved, the governor objected to the issuing of permits by the corps for oil pipelines that would serve the Weeks Island and West Hackberry sites. Not only did the corps delay issuing these permits, it also placed a moratorium on the issuance of permits for other SPR activities in Louisiana. Finally, in February 1978 the governor removed his objection after signing an "understanding" with DOE Deputy Secretary O'Leary that made a number of promises related to implementation of the SPR program and other issues. For example, DOE promised to take no actions that would cause loss of jobs in local industries. (Hence, the delayed utilization of Weeks Island and the loss of storage capacity at Bayou Choctaw.) DOE also promised not to store nuclear wastes in Louisiana without permission from the state. Additionally, the understanding required DOE to locate the SPR Project Office in Louisiana (a decision already made by DOE). As the governor later explained, "If the federal government is going to pour money down a rat hole, I would just as soon it be a rat hole in Louisiana."[6]

The understanding did not resolve all problems with the State of Louisiana. The Department of Natural Resources objected for environmental reasons to the location of a water intake structure on a lake adjacent to the West Hackberry site. Because an independent assessment by scientists from Texas A & M University supported the position that the structure would cause no significant environmental impacts, the SPR Office assumed that the objection would eventually be withdrawn and therefore continued planning for the structure. It later became clear, however, that the Department of Natural Resources would not withdraw its objection. To avoid further delay, the SPR Office decided to locate the intake structure on the Intracoastal Waterway more than four miles away from the West Hackberry site. The objection and the SPR Office reaction to it resulted in the permanent drawdown system at West Hackberry being available in October rather than March 1979. The SPR Office has estimated that the relocation involved additional costs of over $8.8 million.[7]

Whereas the permitting process of the Corps of Engineers is extremely sensitive to the concerns of state governments, the permitting process of EPA is sensitive to the concerns of local special interest groups. By threatening to request adjudicatory hearings, such groups can gain considerable influence, particularly when permits are needed on a timely basis for schedule adherence. The SPR Office discovered how strong this influence can be in its attempts to obtain an EPA permit for the discharge of brine from the Bryan Mound site.

The SPR Office first submitted in April 1977 an application to EPA for a permit to discharge brine into the Gulf of Mexico at a point five miles from shore. The staff expected quick acceptance of the application because it was based on the recommendations of a panel of experts convened by the National Oceanic and Atmospheric Administration (NOAA) for the SPR Office. At the request of the SPR Office, NOAA initiated baseline biological studies of the proposed five-mile disposal site and an alternative twelve-and-one-half-mile disposal site. These studies found no significant differences in biologic productivity between the two sites. NOAA data also supported the SPR Office position that environmental impacts beyond the near field, within a radius of one hundred feet from the brine diffuser, would be minimal. Subsequently, a consultant retained by EPA to review the analysis concurred with the SPR Office conclusion that the five-mile system was environmentally acceptable.

However, the regional EPA Office was unwilling to issue a discharge permit for the five-mile system in the face of vocal opposition from local commercial fishermen and environmental groups. The regional office was willing to issue a permit for the twelve-and-one-half-mile system. In September 1978 the SPR Office decided to accept the permit for this more expensive system in order to avoid further delay. The regional office then required the SPR Office to work with the local groups, which threatened to request adjudicatory hearings, on the development of a monitoring plan for the brine disposal system. Agreement on a monitoring plan was not reached until August 1979, twenty-eight months after submission of the original permit application. The SPR Office has estimated that construction and operations costs would have been over $28 million lower if a permit for the five-mile system had been issued in November 1977 as initially anticipated.[8] More significantly, an additional 20 million barrels of oil could have been put into storage during 1978, before purchases were suspended because of Iranian production reductions, at a delivered price of about $15 per barrel. When purchases were resumed in late 1980, the delivered price was more than $35 per barrel. Thus nominal oil acquisition costs would probably have been $400 million lower without the Bryan Mound permit delay.

6. Creation of DOE

DOE has been controversial since its creation on October 1, 1977. Many have criticized the way DOE has been managed. Others have questioned the wisdom of its establishment and continued existence. Whether or not the creation of DOE should be viewed as a mistake from the broad perspective of energy policy and government organization, it was detrimental to implementation of the SPR program. In addition to the problems caused by the switch from the FEA procurement system to that of ERDA, severe personnel problems and a loss of organizational effectiveness accompanied the creation of DOE.

Because its constituent agencies had more employees than its authorized personnel ceiling, DOE was immediately subjected to an OMB-imposed hiring freeze. As a result, the SPR Office was unable to hire already recruited persons with critically needed engineering and managerial skills. Although the unfilled positions were lost, several requests for hiring permission were eventually granted so that new employees were gradually added to the staff. However, preparation of documentation for the requests diverted already scarce managerial effort from the ongoing implementation effort. The pinch became so severe that most of the strategic planning staff were reassigned to managerial duties.

The manpower problem was aggravated by the shifting of personnel to staff the new organizational structure. Thomas Noel, until then assistant administrator for the SPR under FEA, was named acting assistant secretary for resource applications. He selected more than a dozen SPR Office personnel for his staff, including Robert Davies who had been the first SPR office director. Carlyle Hystad, the acting SPR office director for the nine months following Noel, faced the problem of replacing the lost staff. The advantage of reporting to Noel, who was already thoroughly familiar with the program, was lost when Noel resigned after someone else was selected to fill the assistant secretary position permanently. Meanwhile, Hystad's acting status, which lasted until July 1978, introduced uncertainty about the future management of the SPR Office, which made dealing with the personnel problems even more difficult.

The change in organizational position of the SPR Office became an important factor after Noel left. Under FEA, the SPR Office was directly under the administrator. Under DOE, the director of the SPR Office reported to the deputy assistant secretary for resource applications. Above him was the assistant secretary for resource applications, the under secretary, the deputy secretary, and finally the secretary. Consequently, it was necessary to go through more administrative levels than before to obtain critical decisions and to gain leverage in dealing with other parts of the organization. It is not surprising that it began to take more time and effort for the SPR Office director to get things done within DOE. Even after the SPR Office was placed directly under the under secretary in December 1978, obtaining cooperation within the DOE bureaucracy remained difficult. As Jay R. Brill wrote shortly after retiring as deputy under secretary for the SPR:

> Internally, we don't seem to have achieved the corporate weldment we need, and I have encountered "fiefdoms" that don't have a sense of mission and corporate commitment. A significant amount of my time was consumed in "pushing wet noodles," having to follow up on many commitments, or by having to set up summit meetings or make phone calls to people at the Secretarial Officer level to get support.[9]

Because administrators above the SPR Office all dealt with many other programs that had to be funded out of the DOE budget, greater attention was given to the cost of the SPR program. Whereas under FEA adhering to the schedule was the dominant factor in decision making, under DOE avoiding cost overruns became at least as important. This was especially the case with the first under secretary, who did not accept the argument that increasing world oil prices often justified short-run cost overruns that permitted earlier oil purchases.

The creation of DOE placed an educational burden on the SPR Office. In addition to the hierarchy, key persons in the legal, financial, and administrative offices had to be informed about the program. The presentation of monthly briefings to top DOE managers placed demands on the already overextended staff that did not seem to yield payoffs in terms of greater understanding of implementation problems and the support needed to deal with them.

7. Vulnerability through Failure

In the latter half of 1978, revelations of schedule slippage and cost overruns led to a general perception that the SPR program was being mismanaged. The SPR Office found itself subject to an increasing number of congressional oversight hearings, General Accounting Office studies, and investigations by the DOE inspector general. Scarce staff resources were consumed in cooperating with these efforts. In most cases the outside investigators focused on second-guessing decisions that had already been made. Consequently, their efforts rarely made specific contributions to improved management and may actually have hindered general improvement by slowing the development of an effective management infrastructure that was already underway.

The reputation of the SPR Office for mismanagement encouraged OMB to delay funding for program expansion. Although the Carter administration was publicly committed to expansion of the SPR to one billion barrels, OMB wished to limit the size to 500 million barrels. In the 1977 and 1978 budget cycles, OMB was able to cut funds for planning for the fourth 250 million barrels of capacity. The low credibility of the SPR program encouraged OMB to cut funds for implementation of the third 250 million barrels in the 1979 budget cycle as well. If the SPR program had had a better track record, the DOE leadership might have been more willing to appeal the cuts to the president.

A persistent reputation for mismanagement even after considerable progress had been made seriously hurt morale and led to timidity in dealing with the external environment. For example, throughout 1979 the SPR program was ridiculed in the press for not having the capability to draw down immediately the oil already in storage. The ridicule was undeserved. Both at the time it was made and in retrospect, the decision to delay installation of the permanent drawdown system until fill was well under way is fully justified—it doesn't make sense to build a facility before it is likely to be used, especially if doing so would detract from other aspects of the program. Even if the capability had been available in early 1979, it is extremely unlikely that the president would have ordered a drawdown of a reserve as small as 92 million barrels in response to the Iranian production cuts of more than 5 million barrels per day. Nevertheless, neither DOE nor the SPR Office made a serious attempt to counter the unwarranted ridicule.

Revelation and Recovery

By November 1977 the SPR Office knew that the goal of having 250 million barrels in storage by the end of 1978 could not be realized. A recommendation was made to Secretary Schlesinger that up to 80 million barrels of temporary storage be leased in Rotterdam, Netherlands, and in the U.S. so that oil purchase could continue at a rate consistent with the goal. The secretary decided not to act on the recommendation.

In February 1978 the SPR Office concluded that only between 100 million and 125 million barrels would be in storage by the end of the year. Under pressure from the secretary, the SPR Office agreed to present the more optimistic level as its assessment. Congress was informed that only 125 million barrels would likely be in storage by the end of 1978. Several members of Congress, particularly Senator Henry Jackson and Congressman John Dingell, began expressing concern over the slippage.

Believing the slippage was primarily due to poor management, the secretary selected an experienced manager to serve as SPR Office director. Joseph R. DeLuca, a retired general who had been controller of the air force, took over in mid-July of 1978. He took a number of steps to strengthen the management of the program. Most importantly, he established realistic schedules and cost baselines for the development of storage capacity.[10] Phase I, consisting of 248 million barrels of largely existing capacity, would be available by the end of 1980 and Phase II, consisting of 290 million barrels of newly leached capacity at existing sites, would be available by the end of 1986. Expansion beyond 538 million barrels, Phases III and IV, would be accomplished by contracting with the private sector for completed sites. (As discussed in chapter 7, this "turnkey" approach was abandoned in 1979.)

General DeLuca devoted much of his effort to improving the management infrastructure and overall systems design of the program. He resigned for personal reasons in February 1979 and was succeeded first by General Jay R. Brill and later by Harry A. Jones, who served as deputy assistant secretary for the SPR until January 1982. Both Brill and Jones worked to strengthen further the management infrastructure for controlling construction and operations. Although hiring and retaining appropriately skilled personnel continues to be a serious problem, especially in the Project Office, by the end of 1979 the SPR Office reached a level of organizational maturity that enables it to meet routinely its schedule and cost estimates for the development of storage capacity.

Current Status

In September 1980 Phase I facilities were completed, providing 248 million barrels of storage capacity, which can support a drawdown rate of about 1.7 million barrels per day. Leaching of Phase II storage capacity began in March of 1980. It now appears that Phase II will be completed in mid-1986. Between now and then storage capacity will gradually grow to 538 million barrels and drawdown capacity to 3.5 million barrels per day.

By the end of FY 1981 almost $7.75 billion will have been invested in the SPR program (see table 3). By the end of 1986 an additional $175 million will likely be spent on completion of Phase II facilities. Assuming a moderate growth in real oil prices and continuous fill between now and the end of 1986, an estimated $16 billion will be spent on oil to fill the reserve to 538 million barrels.

At the end of 1980 the Carter administration finally decided to seek funds for construction of Phase III facilities. The administration's FY 1982 budget proposal, endorsed by the Reagan administration, requested funds to begin further expansion of the Bryan Mound and West Hackberry sites and development of one new site. Phase III as currently planned would provide an additional 212 million barrels of storage capacity by the end of 1989. During 1981, before the magnitude of the oil "glut" and the federal deficit became apparent, DOE analysts investigated alternatives for accelerating the development of Phase III and providing greater storage capacity by the middle of the decade. It now appears, however, that the Reagan administration has weakened its commitment to full implementation of Phase III.

The SPR contained 107.8 million barrels of oil at the end of 1980. By the end of 1981, it contained about 230 million barrels, 20 million less than the amount that was supposed to be in storage by the end of

Table 3
Strategic Petroleum Reserve Appropriations (in thousands of dollars)

FISCAL YEAR	*PETROLEUM ACQUISITION AND TRANSPORTATION*	*STORAGE FACILITIES DEVELOPMENT AND OPERATIONS*	*PLANNING*	*PROGRAM[2] DIRECTION*	*TOTAL*
1976	——	300,000	12,000	1,975	313,975
1977	440,000	——	4,000	3,824	447,824
1978	2,703,469	463,933	7,915	6,789	3,182,106
1979	2,885,670	103,290	12,200	5,911	3,007,071
Reprogramming During 1979	-529,214	+529,214			——
Reprogramming During 1980					
#1	-20,391		+12,000	+8,391	——
#2	-1,881			+1,881	——
Transfer in FY 1980[1]				190	190
FY 1980 Rescission	-2,000,000				-2,000,000
FY 1981 Appropriations	1,383,282	82,834	8,000	10,884	1,485,000
FY 1981 Supplemental	1,305,000			507	1,305,507
TOTAL APPROPRIATIONS	6,165,935	1,479,271	56,115	40,352	7,741,673

SOURCE: U.S. Department of Energy, "Strategic Petroleum Reserve: Annual Report," February 16, 1981. p. 18.

[1] Represents funds transferred to the Strategic Petroleum Reserve in FY 1980 to accompany a transfer of five positions.

[2] Excludes funds appropriated to other DOE accounts but used to finance aspects of SPR Program Direction.

1978. Because 91.7 million barrels of the oil were purchased before the dramatic price rises following the Iranian revolution, the market value of oil in storage at the end of 1981 still exceeded the total undiscounted program costs by several hundred million dollars. However, this comparison will become less favorable as oil prices fall. Of greater significance, the SPR is only now beginning to approach a size where it can potentially play an important role in reducing the adverse impacts of oil market disruptions.

SPR Implementation in Perspective

Government programs are often created as political responses to publicly perceived problems. As long as the precipitating problem remains salient, political pressure for visible results is likely to remain strong. In fact, the greater the public concern, the stronger the pressure for fast action is likely to be. Perceived threats to national security will generally have the greatest saliency. Further, such threats may elicit innovative programmatic responses—a space agency to put satellites into orbit, a project to develop ballistic missile-firing submarines, or a large-scale oil storage program. What does the SPR experience suggest about the problems likely to be encountered during attempts to expedite the implementation of innovative programs?

First, the SPR experience shows the risks involved in not following the natural progression of implementation stages from strategic planning, to systems design, to management infrastructure development, and finally to project execution. Congress and the Ford administration, recognizing the vulnerability of the U.S. to oil supply disruptions, imposed ambitious timetables on the program. The one-year period for development of the SPR plan forced the SPR Office to fill most of the initially available personnel slots with persons who could contribute to strategic planning. The ESR deadline required the SPR Office to begin critical project execution tasks before persons with specialized engineering and managerial expertise could be added in adequate numbers to the strategic planning staff. It is not surprising that errors were made in design and procurement. If the SPR Office had been allowed to expand steadily its staff as originally planned, it is likely that most of these errors could have been absorbed as cost overruns without significant schedule slippage. Instead, the personnel ceiling imposed at the creation of DOE, coupled with impossibility of replacing the strategic planners with persons with critically needed skills, doomed the program to both cost overruns and schedule slippage. In general, unless steps are taken to ensure the availability of compensating resources on a timely basis, forcing an unnatural progression of the implementation stages probably will not accelerate and may actually slow program completion.

Second, the SPR experience suggests that importance of giving careful attention to the organizational environment of new programs. Before the creation of DOE, the SPR director reported directly to the FEA administrator. This close contact enabled the director to obtain quickly critical resources and decisions. The SPR was recognized within FEA as having high priority so that competing bureaucratic interests did not greatly interfere with implementation. Once DOE was created, however, the SPR became one of many programs conpeting for scarce resources including the attention of the secretary. The DOE secretary appeared to believe—at least as strongly as the FEA administrator had—in the urgency of the SPR program, but his line of communication with the SPR director was through a multilevel bureaucracy with competing programmatic interests. Requests for waivers from the agency-wide hiring freeze and for relief from the procurement logjam had to be made through an under secretary who placed primary emphasis on minimization of the program's budgetary costs. The request for presidential exemption from environmental permit requirements was stopped by an assistant secretary for policy and evaluation who placed a low priority on contingency planning programs such as the SPR. The FEA experience suggests that the SPR program would have been more successful in securing needed personnel, expediting procurement, and obtaining relief from environmental regulations if its director had dealt directly with the secretary. It seems reasonable to recommend that programs with particularly high priorities be placed directly under department secretaries during their gestation.

Finally, the SPR experience suggests that external oversight is unlikely to play a constructive role during the early stages of implementation when the program character is largely determined. It was only after significant schedule slippage became apparent and DOE had already begun to correct the most obvious implementation errors that OMB, GAO, the DOE inspector general, and congressional committees began looking closely at the management of the SPR program. Oversight usually took the form of second-guessing decisions that had already been made; responding to the oversight placed additional demands on the already overstrained managerial resources of the program. In general, the prospects for constructive oversight of the implementation process are limited because the staffs of the oversight bodies usually cannot accumulate information and expertise as quickly as the program staffs. Further, the attention to the mundane but important details of implementation that is needed for constructive oversight is unlikely to be forthcoming from congressmen seeking publicity, OMB analysts seeking budget reductions, or inspector generals seeking fraud and waste.

Notes

1. My discussion of the implementation of the SPR program is based largely on interviews with Robert L. Davis, Carlyle E. Hystad, Joseph R. DeLuca, Jay R. Brill, Harry A. Jones, Lawrence A. Pettis, Howard Borgstrom, and other program participants. Useful SPR Office documents, currently available through DOE, include the annual reports and Program Stewardship Reports No. 1 and No. 2. Relevant publications include: U.S. Congress, House, "Strategic Petroleum Reserve: Oil Supply and Construction Problems," Hearing before the Subcommittee on Energy and Power of the Committee on Interstate and Foreign Commerce, 96th Cong. 1st sess., September 10, 1979; U.S. Congress, House, "Strategic Petroleum Reserve: Reprogramming of Funds," Hearing before the Subcommittee on Energy and Power of the Committee on Interstate and Foreign Commerce, 95th Cong., 2d sess., December 18, 1978; W. A. Bachman, "Problems Plague U. S. Crude Storage Program," *Oil and Gas Journal*, August 6, 1979, pp. 49-53; Robert G. Lawson, "Strategic Petroleum Reserve Construction Ends First Phase," *Oil and Gas Journal*, July 21, 1980, pp. 47-53; "Special Issue: Strategic Petroleum Reserve," *Journal of the Federation of American Scientists*, 33, no. 9 (November 1980): 1-7; and Ann Pelham, "Energy Department Trying to Work Out Problems of Costly Oil Storage Program," *Congressional Quarterly Weekly Report*, February 3, 1979, pp. 204-5.

2. Strategic Petroleum Reserve Office, "Strategic Petroleum Reserve Plan Amendment No. 1: Acceleration of the Development Schedule," Energy Action No. 12, February 16, 1977.

3. On September 21, 1978, an explosion and fire occurred at the West Hackberry site that killed one worker, seriously injured another, and resulted in an estimated property loss of $12 million. The accident investigation

committee convened by DOE concluded that the cause of the accident "was inadequate attention to critical safety problems, procedures, and emergency response capability." Department of Energy, "Report on the Explosion, Fire, and Oil Spill Resulting in One Fatality and Injury on September 21, 1978, at Well 6 of Cavern 6 at the West Hackberry, Louisiana, Oil Storage Site of the Strategic Petroleum Reserve," DOE/EV-0032, November 1978 (Washington, D.C.: U.S. Government Printing Office, 1978), p. 3.

4. As quoted in Ann Pelham, "Energy Department Trying to Work Out Problems," p. 205.

5. The secretary of energy might have obtained presidential exemptions from several of the environmental regulations that slowed implementation. SPR requests for exemptions were not enthusiastically supported by other offices in DOE, including the Policy and Evaluation Office. Memorandum to George McIsaac, assistant secretary for Resource Applications, from Alvin L. Alm, assistant secretary for Policy and Evaluation, "Action Memorandum Concerning Exempting the SPR Program from Certain Aspects of Environmental Permitting Process," May 17, 1978.

6. Telephone interview with former Governor Edwin W. Edwards, March 2, 1981.

7. Strategic Petroleum Reserve Office, "Impacts of Regulation on the Strategic Petroleum Reserve: A Selective Analysis," October 15, 1979, p. 12. Also available in U.S. Congress, House, "Strategic Petroleum Reserve: Oil Supply and Construction Problems," Hearing before the Subcommittee on Energy and Power of the Committee on Interstate and Foreign Commerce, 96th Cong., 1st sess., September 10, 1979, pp. 76-95.

8. Strategic Petroleum Reserve Office, "Impacts of Regulation on the Strategic Petroleum Reserve: A Selective Analysis," October 5, 1979, p. 11.

9. Letter to under secretary John M. Deutch from Jay R. Brill, October 19, 1979, p. 2. In U.S. Congress, House, "Filling the Strategic Petroleum Reserve: Oversight; and H.R. 7252: Use of the Naval Petroleum Reserves," Hearing before the Subcommittee on Energy and Power of the Committee on Interstate and Foreign Commerce, 96th Cong., 2d sess., April 25, May 21, and September 15, 1980, pp. 69-87.

10. Strategic Petroleum Reserve Office, "Program Stewardship Report No. 1: SPR Baseline," November 3, 1978.

4

The Will and Wherewithal to Fill

Prior to 1979, the strategic petroleum reserve program fell behind schedule due to technical limitations on the rate at which oil could be put into storage. During 1979 and 1980 the program fell even further behind as a result of policy decisions by the Carter administration not to purchase oil to fill available storage capacity. The administration drew little criticism for withdrawing from the "tight" world oil market that prevailed immediately after the Iranian revolution. As market conditions steadily improved during the early months of 1980, however, the reluctance of the administration to resume purchasing became increasingly controversial. Congress finally forced the administration to resume oil acquisition through a provision of the Energy Security Act of 1980.

Upon taking office, the Reagan Administration moved to accelerate acquisition of oil for the SPR. Supplemental funds were secured from Congress to more than double the FY 1981 acquisition rate planned by the Carter administration. The new administration also requested $3.68 billion for continuing acquisition at a high rate in FY 1982. Amid the massive cuts being made in the FY 1982 budget, it is not surprising that many congressmen were anxious to remove such a large expenditure from the budget. Because congressional support for rapidly filling the SPR remained strong, attention was focused on alternatives to the use of direct budgetary allocations for financing oil purchases. Although several substantive proposals appeared, they were rejected in favor of a largely cosmetic change in the way expenditures on oil are counted.

This chapter attempts to answer several questions related to acquisition of oil for the SPR: Why was the Carter administration reluctant to resume oil purchases? What does the fill controversy suggest for the feasibility of implementing economically "optimal" acquisition strategies in the future? What are the advantages and disadvantages of various financing options?

The Fill Controversy

From the middle of 1974 to the latter part of 1978 the world oil market was relatively stable. Although the nominal prices of most grades of crude oil increased slightly, real prices (nominal prices adjusted for changes in the general price level) actually declined.[1] The four-year period of stability ended with the cutbacks in production that accompanied the Iranian revolution. In July 1978 Iranian crude oil production averaged 5.8 million barrels per day. By January 1979 Iranian production fell to less than 0.5 million barrels per day. Spot prices, which reflect day-to-day balances in supply and demand, jumped about four dollars per barrel in the first quarter of 1979 and then more than doubled by the end of the year. Longer-term contract prices followed, also more than doubling their 1978 levels by the first quarter of 1980.[2] The weighted average international price of crude oil rose from $13.80 in January of 1979 to $24.84 in January of 1980.[3]

By the end of 1978 there were 65.9 million barrels in the SPR, and contracts arranged for the delivery of an additional 25.8 million barrels by August of 1979. The Defense Fuel Supply Center (DFSC), the oil purchasing agent for the SPR, continued to issue procurement solicitations in the early months of 1979, but because firms were simultaneously attempting to build their inventories as a hedge against an extended loss of Iranian production, no acceptable bids were received. In May DOE asked the DFSC to suspend further solicitations.

The worldwide accumulation of stocks was quite impressive, especially in light of falling consumption, which reduces the levels of product working stocks needed, and high interest rates, which make holding stocks more expensive. In the first quarter of 1979, world stocks of crude oil and refined products stood at about 4.3 billion barrels, close to the first quarter average for the preceding three years.[4] By the first quarter of 1980, stocks reached a record of 5.1 billion barrels.[5] By mid-1979 it became clear that the worldwide scramble for stocks was contributing to the dramatic rises in spot prices, which many believed would ultimately be followed by higher long-term contract prices. Consequently, in June the U.S. and its major allies agreed at the Tokyo Economic Summit not to buy oil for national reserves if it would place undue pressure on oil prices in the world market.[6]

As early as July of 1979 some observers of the world oil market were urging a resumption of purchases. In a memorandum to Secretary Schlesinger, Under Secretary John Deutch recommended that the DFSC be instructed to resume purchases at a rate of 200,000 barrels per day.[7] He also recommended the development of government-to-government contracts for providing a steady fill rate over the long run. The secretary did not act on the recommendation before his resignation in August.

In the latter months of 1979, market conditions began to improve. Although long-term contract prices were still rising, spot prices began to fall and private sector stock accumulation was continuing. The under secretary assembled an informal interagency committee consisting of representatives from OMB and the Departments of State and the Treasury. By February 1980 the committee reached a tenuous agreement to recommend that oil obtained from the Naval Petroleum Reserve be used for filling the SPR.[8] At a February 8 hearing before the House Subcommittee on Energy and Power, Charles W. Duncan, Jr., who had become secretary of energy the previous August, suggested that a decision to resume filling the SPR would be made by the end of the month.[9]

Secretary Duncan announced a very different policy upon his return from talks with Saudi Arabian leaders in early March. He expected that the administration would not resume filling the SPR until June 1981, a year later than anticipated in the budget requests of two months earlier. What factors explain this shift in policy? The secretary himself described the decision as the result of a "complex mix of political, national security, international, and economic considerations."[10] Fundamental to these considerations were perceptions about the impact of resuming purchases on the world market.

The secretary feared that the resumption of purchases would lead to higher oil prices, particularly if Saudi Arabia decided to reduce its production in retaliation. The higher prices would cause economic losses in the oil importing nations, straining relations with our allies and worsening the administration's record in an election year. The White House staff appeared to share these fears. Furthermore, some believed that risks should not be taken as long as private sector inventories continued to expand.[11]

We can investigate the rationality of these fears by attempting to answer three questions: What would have been the market impact of resuming purchases if Saudi Arabia did not retaliate? What would have been the impact if Saudi Arabia did retaliate? How likely was it that Saudi Arabia would retaliate?

Beginning in January 1980 if not earlier, it is likely that SPR purchases would not have contributed to noticeably higher prices. Although long-term contract prices were continuing to rise toward spot market prices, the latter were beginning to fall. Higher prices, a mild winter in the Northern Hemisphere, and a sluggish world economy were leading to reduced consumption of petroleum products. World stocks of crude oil and petroleum products, already at record levels in the first quarter of 1980, were growing at a rate of over one million barrels per day, reaching an all-time record of 5.58 billion barrels in the third quarter of 1980.[12] Several hundred million barrels more than normal were being held in high-cost tanker storage. In order to protect future lifting-rights, firms were not cutting back their production in the face of falling consumption.[13] Many firms would have been willing to sell oil to the SPR out of their short-run stock accumulations. In fact, firms began approaching the SPR Office with offers to sell. Such transfers from excess private stocks to the SPR would not have put upward pressure on price.

Concurrent reductions in OPEC production suggest that SPR purchases could have been made without increasing prices or reducing private stock levels. Beginning in January 1980, OPEC hardliners began shutting-in production rather than lowering prices in the face of falling demand. In March 1980 *Petroleum Intelligence*

Weekly reported, "Softening demand, more than planned production cutbacks, seems to be the reason for a sharp 1.2 million barrel daily drop in OPEC's January crude oil output."[14] Additional production cuts were made during the first half of 1980. In March Venezuela cut production about 200,000 barrels per day in response to the soft market.[15] Although Kuwait had long-term plans to reduce production, many believed that as much as 200,000 barrels per day of its more than 500,000 barrels per day production reduction in April was due to short-term market conditions.[16] Algeria cut back production 150,000 barrels per day in April, and Nigeria began cutting back production in July.[17] By August 1980, prior to outbreak of the war between Iran and Iraq, OPEC production was reduced to 27.0 million barrels per day from the 1979 average of 30.9 million barrels; excluding Iranian production, the drop was from 27.7 million barrels per day to 25.4 million barrels per day.[18] It is reasonable to assume that as much as 500,000 barrels per day of this amount was shut-in because buyers could not be found at the established prices. If this assumption is correct, it is likely that resumption in early 1980 of acquisition at a rate less than 500,000 barrels per day, the maximum feasible SPR fill rate, would have been met by deferral of production reductions by OPEC members rather than price rises. In retrospect it is hard to believe that the 100,000 to 200,000 barrel per day rate being considered by the administration would have had a significant impact on prices.

If Saudi Arabia cut back production in retaliation, the decision to resume filling the SPR very well might have resulted in higher prices. The determining factor would have been the size of the Saudi cutback, the behavior of private stockpilers, and the response of other OPEC members. If Saudi Arabia reduced production by more than one million barrels per day and other OPEC members continued to shut-in capacity at the rate they actually did, substantially higher prices could have resulted. For example, assuming each 10 percent increase in price leads to a short-run reduction in world consumption of 1 percent (that is, a short-run world price elasticity of demand of minus 0.1), a halt in private sector inventory accumulations, and the historical pattern of production reductions by other OPEC members, a 2 million barrel per day Saudi production cut would have pushed prices about 20 percent above the price path that actually occurred.

Because the higher prices would have had to be paid on all imports, the United States and its major allies would have suffered large economic loses.

Because it was likely that firms would not continue to expand their stocks beyond record levels (and even possible that other OPEC members would delay their production cuts), a Saudi cutback of one million barrels per day probably would not have resulted in higher prices. In fact, analysis conducted by the Policy and Evaluation Office of DOE in March 1980, which concluded that immediately resuming SPR acquisitions at a high rate offered expected benefits far in excess of expected costs, assumed that the Saudis would decrease production from 9.5 million to 8.5 million barrels per day after the June meeting of OPEC.[19]

Alternatively, if it was assumed that the Saudis would not cut production if the United States did not resume filling the SPR, the analysis would have to be modified to include the expected costs of lower levels of private sector inventories resulting from the decision to fill. As it turned out, the drawdown of these private stocks proved valuable in moderating the price rises resulting from production losses caused by the war between Iran and Iraq in the latter part of 1980. On the other hand, the possibility that the Saudis would not cut production in response to the decision to fill would also have to be taken into account. On balance, the conclusion of the analysis that immediate resumption of purchases offered expected net benefits probably would have remained unchanged.

Assessing the likelihood that the Saudis would have retaliated is particularly difficult. Although it is possible to posit Saudi national goals relevant to the issue of retaliation, it is impossible to know with certainty how they would have been reconciled within the Saudi political process in early 1980. Nevertheless, a review of likely Saudi goals provides a good starting point.

Saudi Arabia is blessed with immense oil reserves. At current rates of production, Saudi proven reserves would last more than forty years as compared to only about twenty-five years for the OPEC "price hawks" (Algeria, Libya and Nigeria) and the rest of the world. Because the Saudi economy is small and therefore can absorb only limited amounts of investment from current revenues, the Saudis are relatively more concerned about maintaining the real value of oil prices in the long run than other major producers with smaller reserves and larger economies. Higher current prices will stimulate investments in more energy efficient capital stocks and in alternative energy sources that will lower demand for oil in the future, contributing to lower long-run prices. Consequently, it is in the long-run interest of the Saudis to keep prices from rising too quickly.

Saudi production decisions during 1980 were consistent with the goal of short-run price moderation. In an attempt to force OPEC members to agree to a long-run pricing strategy, the Saudis kept production at 9.5 million barrels per day until October when they increased production to about 10.2 million barrels per day in response to the tightening market caused by the war between Iran and Iraq. Not recognizing this aspect of Saudi self-interest, many administration officials, including the DOE secretary, attributed the high production level to "good will" toward the West and hence felt Saudi retaliation to be very likely if the United States did not continue to court the "good will." While SPR purchases could have made it marginally more difficult for the Saudis to achieve price moderation, effective retaliation would have made it impossible. In fact, a large SPR would actually contribute to the Saudi goal of long-run price stability by reducing the size of price increases during disruptions.

Another Saudi goal is the achievement of a favorable resolution of the conflict between Israel and its Arab neighbors. Pursuit of this goal would call for efforts, perhaps including backing up threats with actual retaliation, to keep the United States from filling the SPR. The more oil in the SPR, the less vulnerable is the United States to threats of production cutbacks and, therefore, the less leverage the Saudis have over U.S. policy toward Israel.

The situation was complicated by Secretary Duncan's consultation with the Saudis on the question. If the Saudis had given their blessing, they would have been open to charges from Arab hard-liners that they were being subservient to the United States. Once the Saudis felt obliged to communicate threats, however indirect, the pressure on them to actually retaliate was increased. Simply beginning to fill the SPR, with as little fanfare as possible, probably would have minimized the likelihood of retaliation.

Although our discussion has been of necessity highly speculative, several conclusions seem justified. First, in the absence of Saudi retaliation, market conditions were extremely favorable for resuming purchases. Second, the expected benefits of resuming purchases probably exceeded the expected costs for Saudi retaliatory cutbacks of up to one million barrels per day. Third, Saudi retaliation was far from certain and, if it occurred, it was unlikely to involve production cuts of the more than one million barrels threatened. Consequently, the wisdom of the decision not to resume purchases in early 1980 is highly questionable.

Of course, it can be argued that it is much easier to make an assessment of market conditions after the fact than at the time the decision was made. This is undoubtedly true. Nevertheless, many observers recognized the favorable conditions at the time. In early February Congressman John Dingell and fellow members of his Subcommittee on Energy and Power began criticizing the administration for not resuming purchases in late 1979 when market conditions began to improve. In April Congressmen Dingell and David Stockman challenged Secretary Duncan to explain to them how market conditions could conceivably be better.[20] After traveling to Saudi Arabia in April, Senator William Bradley concluded that market conditions were favorable for resuming purchases even if the Saudis retaliated.

Senator Bradley led congressional efforts to force the administration to resume filling the SPR. He and Senator Robert Dole were able to add an amendment to the synthetic fuels bill requiring the administration to resume filling the SPR at a rate of at least 100,000 barrels per day. In conference on the bill, Congressman Dingell warned that as the amendment was written the administration could ignore it. He recommended that continued production and sale of oil from the naval petroleum reserves be tied to resumption of oil acquisition for the SPR so that the administration would not be able to reduce its budget deficit by further delay. This approach was prompted by the suspicion, shared by several other members of the Subcommittee on Energy and Power, that the administration's new emphasis on balancing the budget was influencing the acquisition decision.[21]

The administration pleaded with congressional leaders to delay action on the amendment until at least after the June OPEC meeting. Congress agreed, but sent the SPR amendment to the president on June 26 as part of the Energy Security Act (PL96-294), which established the Synthetic Fuels Corporation. The amendment required the president to undertake filling of the SPR at a minimum average rate of 100,000 barrels per day in FY 1981 and subsequent fiscal years unless an SPR drawdown is underway. The president would be required to obtain congressional approval for suspension of purchases for other reasons. If the 100,000 barrel per day average is not achieved, the federal share of oil from the naval petroleum reserves must be used for filling the SPR. The Energy Security Act also directed the president to amend the entitlements program so that in effect purchases for the SPR would be at the lower price of domestically controlled oil.

The controversy flared up again when DOE announced that it would attain a 100,000 barrel per day fill rate by trading the federal share of naval petroleum reserve production for oil suitable for storage in the SPR. The plan was completed within DOE by the end of June and approved by the president in July. Members of the Subcommittee on Energy and Power were annoyed by two aspects of the plan. First, it would provide fill at the minimum mandated rate even though the market continued to be slack. Second, use of the cumbersome swapping procedures rather than the more straightforward route of direct market purchases appeared to be pandering to the Saudis.[22] Congressman Stockman accused the administration of "timidity and abject capitulation" and Congressman Phil Gramm charged that DOE negotiations with OPEC members were based on "a naivete that is virtually unbelievable."[23] Chairman Dingell announced plans to introduce legislation that would transfer responsibility for filling the SPR from DOE to an independent public corporation.

A formal congressional response came in early December with passage of the FY 1981 appropriations bill for the Department of the Interior and related agencies (PL96-514). It included a provision that made appropriations for SPR facilities contingent upon the president seeking to increase the fill rate to at least 300,000 barrels per day or a rate that would fully utilize appropriated funds. Although the provision as written can be largely ignored by the president, it does convey the congressional intent that the fill rate be accelerated.

Preparations for open-market purchases of oil began in the last days of the Carter administration. With the support of the acting DOE under secretary, the director of the SPR Office prepared plans for the Defense Fuel Supply Center to initiate a "continuous open solicitation" that would permit regular purchases of spot cargoes. The new acquisition system, which went into effect under the Reagan administration at the end of January 1981, resulted in an increase in the average fill rate to over 200,000 barrels per day. The Reagan administration requested funding to support a 230,000 barrel per day fill rate during FY 1982. The administration also requested and obtained supplemental funds for FY 1981 to offset the loss of entitlement benefits for SPR oil purchases that resulted from the decontrol of domestically produced oil prices in February 1981. In the future, purchases of oil for the SPR would be at full market prices rather than at controlled domestic prices.

The Financing Question

In March 1981 the Reagan administration proposed a FY 1982 budget for the SPR program of $3.9 billion, $3.68 billion of which would be used for the acquisition of oil. Both the Senate and House Budget Committees moved to reduce the amount requested under the assumption that compensating funds would be raised "off-budget" to permit acquisition of oil at the rate planned by the administration. Although several substantive alternatives received attention within Congress and the administration, SPR program financing was simply moved off-budget in an accounting sense for FY 1982. Expenditures will not be counted as part of the budget even though they are to be financed through federal borrowing in the same way as other federal deficit spending. Moving the SPR off-budget in this way will leave expenditures and borrowing levels unchanged.

The financing question attracted congressional attention at a time when reducing the size of the budget had an unprecedented political saliency. The emergence of a broad consensus within Congress for finding some kind of off-budget financing method undoubtedly was related to desires for at least the appearance of a smaller budget. Nevertheless, there was a general perception that the peculiar nature of SPR purchases justifies off-budget financing. Unlike most other government expenditures, money spent purchasing oil for the SPR buys an asset that is likely to appreciate in value. The program thus has potential for self-financing through borrowing against the expectation of revenues from future sales.

A major source of support for moving SPR oil purchases off-budget came from those who advocated private equity financing.

Although numerous variations are possible, the basic elements of equity financing are: the sale of negotiable certificates denominated in barrels of oil to fund oil purchases; and the redemption of certificates in the future at a value related to the then current price of oil. For example, the most prominent proposal for equity financing, Congressman Gramm's Private Equity Petroleum Reserve Act (H.R. 2304), called for the sale of certificates at a price not less than the average weighted price of crude oil imported into the United States during the quarter preceding the date of issue. Certificates would be redeemable after ten years at the prevailing market price of oil less certain storage and handling costs. During a drawdown, the secretary of energy would call in for early redemption a number of certificates equal to the number of barrels being removed from the SPR.

The equity financing approach has several potentially attractive characteristics.[24] First, it provides an opportunity for persons and firms to buy an asset that will appreciate in value during oil supply disruptions. By holding certificates, those particularly vulnerable to economic losses during severe disruptions can provide themselves with a source of revenue for offsetting such losses. In an investment sense, they would be diversifying their portfolios to reduce risk.

Second, the redemption of certificates during disruptions would reduce macroeconomic losses by helping to stabilize national income. Oil price shocks reduce real national income through increased payments to foreign oil producers. Reductions in national income, along with revenue transfers from consumers to domestic oil producers and to the federal government (through the windfall profits tax), reduce aggregate demand and change its composition, further reducing national income and slowing investment.[25] Under equity financing, SPR drawdowns result in revenue flows directly to the private sector rather than the federal treasury. These flows help reduce declines in aggregate demand and, if certificates are held by those otherwise suffering the greatest economic harm, stablize the composition of aggregate demand.

Third, the widespread distribution of certificates would help create a broader political constituency against the reimposition of price controls. As long as the redemption value of the certificates is tied to the price of oil, certificate holders would have an incentive to oppose government attempts to hold domestic prices below market clearing levels. Opponents of price controls would therefore find this aspect of equity financing attractive.

There are, however, a number of potential problems associated with equity financing. It may be that certificates will command auction prices consistently lower than current market prices so that continued federal subsidization of oil purchases would be needed. Such discounting would occur if the marginal purchaser believed that the expected price of oil will grow slower than the rate of interest. To maintain auction prices sufficiently close to current market prices (so no revenue supplements would be needed), certificates might have to be issued at a slower rate than the government believes to be socially optimal in terms of economic and national security benefits provided by the SPR. Another potential problem, which could be dealt with by restrictions on the ownership and resale of certificates, would be the possibility of manipulation of the certificate market by foreign producers with inside information about the likelihood of future disruptions. Finally, there is a danger that equity financing would displace growth of the private futures markets that have recently developed for certain petroleum products and that appear to be developing for crude oil.[26]

The administration opposed equity financing. OMB Director David Stockman, who, ironically, suggested the equity financing approach in 1978 while a member of the House Subcommittee on Energy and Power, expressed concern that the Gramm proposal might not yield sufficient revenue to finance oil acquisition at the rate planned by the administration.[27] He also conveyed the administration's opposition to the general concept of equity financing embodied in S.998, a bill introduced by Senators James A. McClure and John W. Warner, which guaranteed multi-year funding through a combination of financing sources.

The other major substantive proposal for alternative financing was introduced by Senator Nancy A. Kassebaum in the Strategic Petroleum Reserve Amendments of 1981 (S.707). Her bill would require importers of more than an annual average of 75,000 barrels per day to contribute oil to the SPR at an annual rate of five times their average daily imports during the previous calendar year. The government would pay the importers an annual fee of 10 percent of the purchase price of their contributions for 11 years or until the oil is withdrawn, whichever is sooner. When the oil is withdrawn during a drawdown, firms would receive a per barrel payment equal to the average market price of the three preceding months less the annual fees already paid.

Senator Kassebaum's plan would provide about 25 million barrels per year of oil for the SPR at a substantially reduced cost to the federal government. Total federal expenditures on SPR oil in FY 1982 would be reduced by about one billion dollars. The plan, however, has several weaknesses. The government would be forced to monitor sale prices in order to determine the fee rate. By raising the crude oil price to U.S. refiners, the plan would provide an implicit subsidy for petroleum product imports. The size exemption might encourage the proliferation of a large number of small importers, resulting in uncertain economic consequences. Finally, the rationale for the payment schedule is not clearly developed. Firms would undoubtedly suffer financially if there were no drawdowns and gain if there were. On balance, it is unclear how much the firms would suffer or gain under the plan.

Senator Kassebaum's proposal can be viewed as a type of in-kind import fee. A more straightforward version had been considered the previous year by analysts in the Policy and Evaluation Office of DOE.[28] Under their in-kind import fee, crude oil and petroleum product importers would be required to deliver a fixed percentage of their imports to the SPR. The nominal liability would be in terms of an established percentage of the quantity of each grade of crude imported by the firm. Product importers would have liability for a crude equivalent of their product imports. With the permission of the SPR Office, importers could substitute grades of crude or substitute a cash payment. Delivery would be made during each quarter for the fee liability of the previous quarter. Actual physical delivery likely would be made by a small number of intermediary companies that accumulate liabilities through trades.

A fee rate of 3 percent would yield an average of about 210,000 barrels per day of SPR fill at current levels of imports. Importers would view the fee as a 3 percent ad valorem tariff. At current prices of approximately $35 per barrel, the tariff would be a tax of $1.05 per barrel. Unlike a percentage tax on the value of imports, revenues would not flow through the treasury. Rather, the government would accumulate holdings of oil. During periods when no additional SPR storage capacity was available, the fee would be automatically suspended. When SPR capacity permits, importers would be allowed to bank deliveries in anticipation of future imports. This banking provision would encourage importers to take advantage of low spot market prices during slack market periods and thereby achieve earlier and lower cost SPR fills. Once the fee was put in place, fill would occur without explicit decisions by Congress or the administration.

The in-kind import fee is consistent with two general policy positions. The first is the belief that a tariff on imported oil is a desirable policy tool because the social cost to the United States of the marginal barrel of imported oil exceeds its market price by some "premium" related to increased dependence and vulnerability. Domestic consumers and producers will not make socially efficient decisions unless they see the correct social price. Hence, the in-kind import fee is desirable because it raises the market price of imported oil closer to its social cost.[29]

The second is the belief that the SPR is primarily an insurance policy against the economic losses of disruptions. Price reductions resulting from SPR drawdowns will provide economic benefits to consumers roughly proportional to petroleum product use. Therefore, it is reasonable that consumers pay for this protection through the higher prices caused by the in-kind import fee.

Although the Reagan transition team for DOE expressed some interest in the in-kind import fee, the administration rejected the general concept as well as Senator Kassebaum's specific proposal. The administration opposed the introduction of new taxes and rejected the import "premium" argument. It also tended to view the SPR more as a national security program that should be funded out of general revenues.

While opposing equity financing and in-kind import fees and continuing to state its preferences for on-budget financing, the administration did not oppose the creation of a separate SPR account to be funded through treasury borrowing. Congress eventually adopted this approach for FY 1982, deferring possible adoption of more substantive alternative financing proposals until the next session.

Acquisition Strategies

From an economic point of view, an optimal acquisition strategy should take into account the amount of oil already in the SPR, the current state of the market, and expectations about future states of the market.[30] Generally, during slack market periods, the reserve should be filled at the maximum rate that is technically feasible. During tight market periods, the optimal acquisition decision depends on how much oil is already in the reserve and expectations about future states of the market. Implementation of the optimal strategy requires decision makers to recognize and quickly react to changing market conditions. The 1980 fill controversy suggests the political and institutional limitations to the requisite flexibility.

Congress and the Reagan administration seem committed to establishing a moderate but steady fill rate. Although not optimal from a strictly economic point of view, it may be the best strategy for filling the SPR under existing institutional arrangements. Could a more nearly optimal acquisition srategy be successfully implemented under different institutional arrangements?

One promising approach would be to restructure the SPR program as an independent public corporation with an initial multi-year standing appropriation and the authority to undertake equity financing. Congress and the administration would jointly establish a minimum fill rate that the corporation would be required to meet, perhaps through an in-kind import fee. Additionally, the corporation would have a mandate to accelerate acquisition to the highest technically feasible rate whenever spot cargoes could be purchased at below long-term contract prices. As long as spot prices are below contract prices, purchases are unlikely to put upward pressure on contract prices.

The spot price purchase mandate offers two major advantages over the current acquisition system. First, it would accelerate the acquisition rate during slack market periods, resulting in earlier and less costly SPR fill. Second, it provides a mechanism for the automatic resumption of purchases after acquisition has been suspended. The corporation would continually seek spot cargoes at below contract prices, purchasing them whenever found. The decision to resume acquisition would have a very low visibility and therefore be less likely to be subject to threats or retaliation. Also, if potential retaliators believe the mandate cannot be easily overruled by Congress and the president, they may be less likely to threaten retaliation because they will be more likely to have to follow through on their threats.

Notes

1. For example, consider the nominal and real contract prices for three primary OPEC crude oils:

	MIDEAST LIGHT	*MIDEAST HEAVY*	*AFRICAN LIGHT*
NOMINAL PRICE			
1974, 4th Quarter	$10.40	$10.17	$11.75
1978, 4th Quarter	12.70	12.27	13.87
1980, 1st Quarter	27.17	27.90	34.67
REAL PRICE (1972 $)			
1974, 4th Quarter	$ 8.70	$ 8.50	$ 9.82
1978, 4th Quarter	8.19	7.92	8.95
1980, 1st Quarter	15.87	16.30	20.25

SOURCE: *Petroleum Intelligence Weekly*, October 20, 1980, p. 11.

2. Ibid.

3. U.S. Department of Energy, Energy Information Administration, "Weekly Petroleum Status Report," July 3, 1981, p. 20.

4. U.S. Department of Energy, International Affairs, "International Energy Indicators," June 1981, p. 8.

5. U.S. stocks went from 1.15 billion to 1.34 billion barrels over the same period. U.S. Department of Energy, "International Energy Indicators," p. 8.

6. Tokyo Communique, "Joint Declaration of Tokyo Summit Conference," June 29, 1979. In *Public Papers of the Presidents of the United States, Jimmy Carter, Book II, June 23 to December 31, 1979* (Washington, D.C.: U.S. Government Printing Office, 1980), pp. 1197-1201.

7. Memorandum from under secretary to secretary, "Continued Fill of the Strategic Petroleum Reserve," July 11, 1979. In the U.S., Congress, House, "Filling the Strategic Petroleum Reserve: Oversight; and H.R. 7252: Use of the Naval Petroleum Reserves," Hearings before the Subcommittee on Interstate and Foreign Commerce, 96th Cong. 2d sess., April 25, May 21, and September 15, 1980, pp. 62–68. (Hereafter cited as "Hearings on Filling the Strategic Petroleum Reserve.")

8. Christopher Madison, "How Can We Build an Oil Reserve Without Offending the Saudis?", *National Journal*, June 28, 1980, pp. 1044–49.

9. U.S., Congress, House, "Fiscal Year 1981 Authorization for the Department of Energy and the Federal Regulatory Commission," Hearings before the Subcommittee on Energy and Power of the Committee on Interstate and Foreign Commerce, 96th Cong., 2d sess., February 8, 11, 12, 13, 20, and 29, 1980. p. 82.

10. "Hearings on Filling the Strategic Petroleum Reserve," p. 42.

11. Interview with former National Security Council staff member Ed Fried, June 19, 1981.

12. U.S., Department of Energy, International Affairs, "International Energy Indicators," June 1981, p. 8.

13. For example, there were reports in the trade press that Algeria's decision to reduce contract sales volumes was greeted with relief by companies that had been continuing unprofitable production under the contracts rather than risk loss of future access by reducing liftings. "Algeria Joins Efforts in OPEC to Reduce Oil Surplus," *Petroleum Intelligence Weekly*, March 31, 1980, pp. 3–4.

14. "Softening Demand Causing OPEC Oil Output to Drop," *Petroleum Intelligence Weekly*, March 10, 1980, p. 1.

15. "Venezuela Cutting Its Crude Oil Output but Not Its Ceiling," *Petroleum Intelligence Weekly*, March 24, 1980, p. 1.

16. "Kuwait's Cutback Plan Leaves Scope for Flexibility," *Petroleum Intelligence Weekly*, February 25, 1980, pp. 5–6.

17. "Algeria Joins Efforts in OPEC to Reduce Oil Surplus," *Petroleum Intelligence Weekly*, March 31, 1980, pp. 3–4.

18. U. S., Department of Energy, Energy Information Administration, *Monthly Energy Review*, June 1981, pp. 88–89.

19. U. S., Department of Energy, Office of Policy, Planning, and Analysis Division, Energy Security SPR Files, "Draft Budget Issue Paper," RA Volume IVA1-A40, B1-B21, March 21, 1980.

20. "Hearings on Filling the Strategic Petroleum Reserve," pp. 43–44.

21. For example, Congressman Gramm expressed the view that delaying oil purchases allowed DOE to meet the new budget proposed by the administration in March without "cutting programs that had a clear or more immediate constituency." "Hearings on Filling the Strategic Petroleum Reserve," p. 44. Secretary Duncan emphatically denied that the decision was being influenced by budgetary considerations. Whether or not it influenced the secretary's decision, the prospect of relatively painless budget cuts probably predisposed others in the administration to accept Duncan's assessment of the impact of Saudi reactions to resumption of fill. Later there were reports from former DOE officials that OMB blocked attempts by Duncan to accelerate fill once it was resumed. As reported by Christopher Madison, "The Energy Department at Three—Still Trying to Establish Itself," *National Journal*, October 4, 1980, p. 1649.

22. Abram Chays, special adviser to the secretary on SPR matters, acknowledged in congressional testimony that the belief that the Saudis would find the NPR exchange approach less offensive than open market purchases was a factor in its selection. "Hearings on Filling the Strategic Petroleum Reserve," p. 356.

23. "DOE Slates 100,000 b/d of Oil for SPR by Dec. 1," *Oil and Gas Journal*, September 22, 1980, p. 57.

24. For an excellent discussion of equity financing options for SPR oil see Michael Barron, "Market-oriented Financing of Oil Stockpile Acquisition," Staff Working Paper, Office of Policy, Planning and Analysis, DOE, February 15, 1981. Also see U. S., Congress, Congressional Budget Office, "Financing Options for the Strategic Petroleum Reserve," April 1981; and Philip K. Verleger, Jr., "Let the Market Fill the U. S. Petroleum Reserve," *Wall Street Journal*, April 27, 1981, editorial page.

25. Chapter 5 provides a discussion of the macroeconomic effects of oil price shocks.

26. Roger Vielvoye, "The Futures Market," *Oil and Gas Journal*, February 16, 1981, p. 66.

27. Statement of David A. Stockman, director of the Office of Management and Budget, before the Subcommittee on Fossil and Synthetic Fuels of the Committee on Energy and Commerce, U. S. House of Representatives, May 1, 1981.

28. David L. Weimer, "Routine SPR Acquisitions: The In-Kind Import Tariff and Spot Market Purchase Authority," Office of Oil Staff Paper, Policy and Evaluation Office, DOE, September 10, 1980; and John H. Jennrich, "In-kind Crude Tariff Pushed for the SPR," *Oil and Gas Journal*, December 15, 1980, p. 46.

29. This assumes, of course, that the difference between the true social cost and the market price is greater than the size of the tariff implied by the in-kind import fee. Most estimates of the size of the optimal long-run import tariff are larger than a 3 percent in-kind import fee. For example, see William W. Hogan, "Import Management and Oil Emergencies," in *Energy and Security*, David A. Deese and Joseph S. Nye, eds. (Cambridge, Mass.: Ballinger Publishing Company, 1981), pp. 261–301.

30. Glenn Sweetnam et al,. "An Analysis of Acquisition and Drawdown Strategies for the Strategic Petroleum Reserve," Draft, DOE, Assistant Secretary for Policy and Evaluation, Office of Oil, December 17, 1979.

PART II

HISTORY OF SPR ANALYSIS

5

Basic Analytical Considerations for SPR Size Studies

Should the United States develop a strategic petroleum reserve? If so, how large should it be? Extensive efforts have been made by analysts, both inside and outside government, to answer these questions. Most analyses have been within the general economic framework of comparing the expected social costs and benefits of various SPR sizes. Within this framework, however, there are a number of alternative modeling options for dealing with the measurement of social costs and benefits and the treatment of uncertainty about future events. Even when analysts employ models with similar structures, they can obtain widely divergent results by making different assumptions about future world conditions and how economic and political actors will react to them. The objective of this chapter is to provide, for its own interest and as background for subsequent discussion of the controversy over the appropriate size of the SPR, a simple review of the key analytical issues that must be addressed in determining the optimal stockpile size for the United States.

As is almost always the case with questions of public policy, there are a number of relevant factors in determining the appropriate SPR size that cannot be or at least have not been adequately accounted for within the commonly employed cost/benefit framework. Not only are these factors of potential substantive importance, they also serve as the last refuge for those whose policy positions are challenged by results from the formal economic models. Before limiting our attention to questions related to formal modeling, it is worthwhile to review briefly these excluded factors.

A large stockpile expands the range of options the U. S. can pursue in carrying out foreign policy. Stockpile drawdowns provide a "breathing-space" between the time oil supplies from an exporting region are disrupted and the time the economies of the United States and other net importers feel the full impact of the resulting price increases. During this period diplomatic initiatives can be launched in a less politically volatile domestic environment than would exist if consumers were bearing the full impact of the disruption. In the event the disruption is coincident with or leads to military intervention, the stockpile would enhance military flexibility by reducing the costs to the United States and its allies of temporary damage to production facilities. Although the value of such flexibility is difficult to quantify, it is nevertheless real, especially to a president who must make politically sensitive decisions such as those concerning the resupply of Israel during a Middle East war or the use of military force to aid the government of Saudi Arabia against foreign-supported insurgents.

A large stockpile may deter embargoes and politically motivated reductions in production. The larger the stockpile, the more severe the reduction in supply must be to inflict any particular level of economic costs on importers. As long as exporting countries anticipate that they also will suffer economic losses by reducing production, the existence of a large stockpile will tend to deter the use of supply disruptions for political purposes. Although attempts have been made to derive quantitative estimates of this deterrent effect through the application of game theory, the results cannot be readily integrated within the cost/benefit framework.[1]

Efforts to develop a large stockpile may elicit retaliatory cutbacks in production from nations that anticipate possible future use of disruptions as instruments of their foreign policy. Such retaliatory cutbacks make stockpile development relatively less attractive by increasing the price of oil, not only for stockpile purchases, but for all imports. If the cutbacks are perceived as permanent by other producers and by consumers, the direct costs of the price increase will be offset somewhat by reduced vulnerability to supply disruptions in the future that results from the increased investment in conservation and production induced by higher expected prices. If the cutbacks are in fact permanent, vulnerability of the stockpiling nation may also be lessened because the retaliating nation would have to reduce production by a larger percentage than before the retaliation in order to cause a supply disruption of a particular magnitude. On the one hand, the reduction in vulnerability resulting from a permanent retaliatory cutback argues that less weight should be given to the threats of retaliation in making stockpile development decisions. On the other, it argues for a smaller stockpile because the reduced vulnerability will translate into smaller expected benefits from use of the stockpile. These considerations, coupled with uncertainty over whether any such threats would actually be carried out in light of linkages with other economic and political issues, make it very difficult to deal with retaliation in the cost/benefit framework.

The existence of a large stockpile controlled by the U.S. Government may discourage stockpiling that would otherwise be done by domestic industry. Firms would anticipate that drawdowns from the government reserve during supply disruptions would reduce the market price at which they could sell their stockpiles. Although this price leveling effect will reduce the size of the stockpiles firms will choose to hold, the marginal reduction due to the government stockpile may be small because firms anticipate the possibility that price controls and government allocation of private stockpiles will be reimposed during disruptions large enough to trigger use of the government stockpile. As firms alter their expectations about the likelihood of government intervention in the market and the probable ways the government stockpile will be used, the displacement of private stockpiling could be significant.

A U.S. strategic petroleum reserve also may discourage foreign governments from stockpiling oil. Because all consumers gain from the lower price resulting from drawdown of the U.S. stockpile, other importing nations may take a "free ride" on large U.S. stockpiles rather than developing their own. Additionally, the benefits of the U.S. stockpile will depend on the stockpiling behavior of other nations. Although some analyses have dealt with the displacement questions endogenously, most have done so through exogenous assumptions.[2]

Turning now to the assumptions underlying the formal economic models used to evaluate stockpiling policies, we will focus on three basic questions: What are the consequences of stockpile acquisitions and drawdowns under various market conditions? How do we measure the costs and benefits of these consequences? How do we account for our uncertainty over future market conditions?

Consequences of Stockpiling and Disruptions

An evaluation of stockpiling policies must begin with predictions about the price of oil in future periods. Supply disruptions inflict costs on net importers by raising the price of oil in the world market. Similarily, stockpile acquisitions may inflict costs by raising prices. Stockpile drawdowns during supply disruptions provide benefits by moderating price rises. Before making predictions about the impacts of supply disruptions and stockpile size changes on the world price, it is necessary to make assumptions about the price trajectory likely to prevail in their absence.

One way to derive the price trajectory would be to assume specific supply and demand curves that would change gradually over time. At any particular time, the price that cleared the world market would correspond to the intersection of the supply and demand curves. A somewhat different procedure may prove more appropriate for two reasons.

First, the world market is not fully competitive. In order to construct supply curves independent of demand curves, we must assume that firms act as price takers. That is, they act as if the amounts they individually offer to the market at each stated price are so small that their production decisions will not noticeably influence the price that actually clears the market. At the other extreme, a firm with monopoly power fully realizes that the amount it offers will determine the market clearing price according to the demand curve it faces. Consequently, it is not possible to derive a supply curve for the monopolist that is independent of the demand curve. The same argument applies to oligopolists, who additionally must anticipate that their individual production decisions will influence the production decisions of the others.

It is reasonable to view the world oil market as being dominated by a relatively small number of oligopolists who act collectively to some degree through OPEC. Just prior to the Iran-Iraq war in 1980, the five largest OPEC exporters were supplying about 40 percent, and

Saudi Arabia alone about 20 percent, of free world consumption. Despite the existence of a large number of small producers that do compete as "price takers," it is not reasonable to assume the existence of a well-defined supply curve. Instead, we will envision the oligopolists attempting to establish collectively a price trajectory over time that maximizes the present value of their joint profits.

Second, the theory of exhaustible resources suggests what the profit maximizing price trajectory will look like. Beginning with Harold Hotelling's classic 1931 article, economists have developed a theory for determining the optimal rate of extraction of exhaustible resources.[3] For a competitively provided finite resource with no close substitutes, the difference between price and the costs of extracting an additional unit must grow at the rate of interest for maximization of the present value of profits.[4] When a resource is owned by a monopolist, the difference between marginal revenue and marginal cost of extraction must grow at the rate of interest. When this condition holds, it will be impossible for the monopolist to increase the present value of profits by shifting production of a unit from one period to another. A number of factors can alter the profit maximizing price trajectory: the existence of close substitutes, the impact of current prices on future demand, the existence of a "backstop" technology that permits the production of a perfect substitute good when a threshold price is reached, and the reestimation of the size of available reserves. These factors complicate the application of the basic theory for predicting the price path of oil.

Empirically, the picture is even less clear. The price of oil in real dollars declined between 1950 and 1970.[5] Since 1973 we have witnessed sharp increases in real prices followed by periods of gradual decline. The apparent reversal in the price trend has been attributed by some observers to the shift in property rights from the multinational oil companies to the nations that previously granted them exploration and production concessions.[6] Because the multinationals anticipated the possibility of nationalization, they discounted the value of oil left in the ground much more heavily than the sovereign nations. Their lower discount rate makes the nations more willing to restrict current production in order to increase price.

Although long-run predictions of price are highly uncertain, it is probably reasonable to assume that the major producers in OPEC will maintain sufficient market power over the next decade to achieve a continuation in the overall trend toward higher real prices that began in 1970. Figure 2 illustrates this assumption. The price in real dollars, in the absence of supply disruptions, is shown as growing from current price, P_O to the backstop price, P_B, at a rate somewhat less than the real rate of interest. Once the backstop price is reached at t_B, it is assumed that there is no longer an upward trend in real price because substitutes are available.

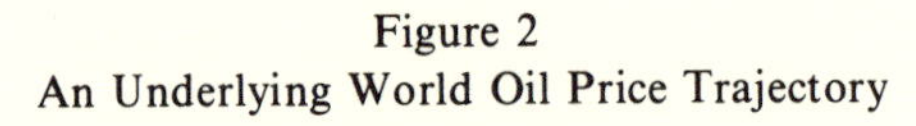

Figure 2
An Underlying World Oil Price Trajectory

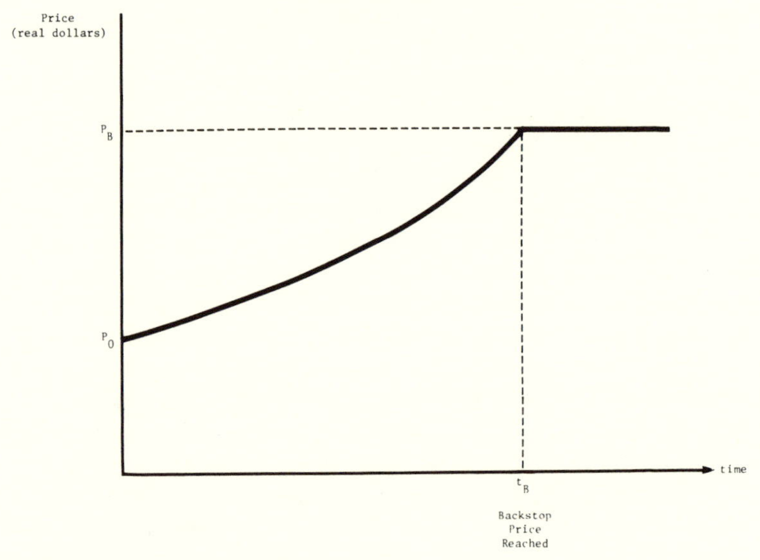

Stockpile purchases may raise the world price above the long-run trend. The size of the price increase will depend upon the response of producers to the temporary increase in demand. At one extreme we can envision a tight market in which suppliers will not provide more oil at the prevailing price in the short run. Buying oil for the stockpile increases demand, but prices must rise for suppliers to be willing to meet this demand. The size of the price increase depends upon the price elasticity of demand, which describes the extent to which consumers cut back on purchases as price goes up.[7] In a slack market, on the other hand, it is assumed that some of the producers are willing to supply additional amounts at the current price, and as a result, that price does not actually increase. We can also imagine intermediate supply response cases where a fraction of the acquisition amount is offset by increased production from the major producers and the competitive fringe.

In figure 3, P_T (A_1, t_1) represents the world price when an acquisition of size A_1 is made for the stockpile during the tight market period between t_2 and t_3. In contrast, the acquisition between t_1 and t_2 is assumed to take advantage of a slack market so that price does not rise above the underlying trend. When oil is purchased for the stockpile during a tight market, all consumers pay the higher price that results. Obviously, the cost of accumulating the stockpile will be lower during slack market periods.

Figure 3
Basic World Oil Price Assumptions

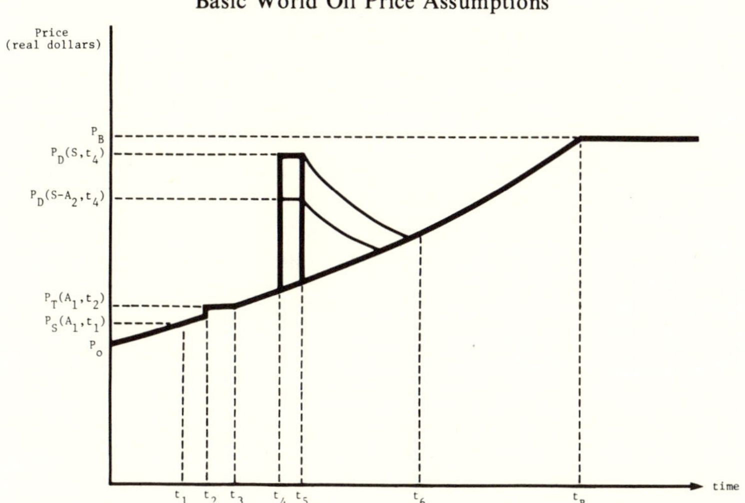

During a supply disruption, a quantity of oil is temporarily removed from the market. The world price will rise to a level at which the amount demanded at the new price equals the available supply. P_D (S, t_4) in Figure 3 represents the market clearing price during a disruption of size S lasting from t_4 to t_5. As shown, the real price remains constant during this period. A more likely pattern would be a decline in real price after the initial sharp price rise as the firms in the competitive fringe expand production in response to the higher prices and consumers have an opportunity to alter their behavior. In a very severe disruption, there also may be macroeconomic effects that shift the demand curve for oil to the left (less demanded at each price) and therefore reduce the world price. Nevertheless, for modeling purposes it is usually sufficient simply to assume that some higher price is reached, which is viewed as an average over the period of the disruption.

What happens to the price when the disruption ends and the previously curtailed supply is returned to the market? Usually it is assumed that the real price returns to the underlying trend that prevailed prior to the disruption. Although conceptually justifiable, the sudden drop in real prices does not fit well with our experiences after the Arab oil embargo. Instead, it might be assumed that the nation controlling the curtailed supply would return it to the market at a rate that would prevent the nominal price from falling. If the nation controlling the curtailed supply does not moderate its resumption of production, other producers might restrict their production to support the nominal price. Even if the nominal price was maintained, the increase in general price levels would gradually reduce the real price of oil. In figure 3 the decline in real price between t_5 and t_6 represents a "nominal price ratchet." At t_6 nominal prices again begin to rise so that real price returns to the underlying trajectory.

Drawdown of the stockpile during a disruption mitigates the price increase by reducing the size of the supply shortfall. For a disruption of size, S, instead of the price rising to P_D (S, t_4) it will only rise to P_D (S-A_2, t_4) where A_2 is the volume of the drawdown. If a price ratchet is assumed, the drawdown reduces the world price not only during the disruption, but also in the following period as the price returns to the underlying price trajectory.

Costs and Benefits of Stockpiling

This section discusses how we can measure the social costs of building a stockpile and the social benefits of using it. Although measuring social costs is relatively straightforward, there is no fully satisfactory method to measure the benefits of stockpile drawdowns during supply disruptions. We demonstrate this conclusion below by describing two methodologies analysts use in estimating such benefits: a microeconomic approach, which estimates changes in social welfare, and a macroeconomic approach, which estimates changes in gross national product.

Estimating Costs

Measuring the social costs of accumulating and holding the stockpile is relatively straightforward. Engineering data can be used to estimate the dollar costs of building and maintaining storage facilities. The social costs of purchasing oil during a slack market is the amount purchased times the world oil price, which represents the marginal cost to the U.S. of oil use.[8] The social costs of oil purchases in tight markets involve two components. The direct cost of the purchase is the new world price times the quantity purchased. The indirect cost is the social surplus (the sum of consumer and producer surpluses) loss associated with the price increase. In figure 4, D(P) is the U.S. demand curve for oil and S(P) is the U.S. domestic

Figure 4
Microeconomic Costs of Acquisitions and Disruptions

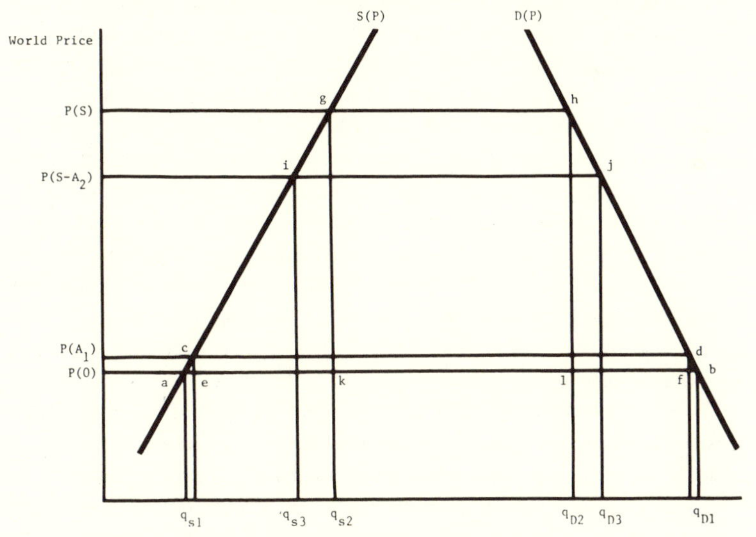

supply curve. If the purchase of stocks of size A_1 raises the world price from P(O) to P (A_1), the social surplus loss is given by the area of the trapezoid, abdc. The area of rectangle cdfe represents the increased payments made to foreign producers. The area of triangle dbf represents the dead weight loss to consumers, and the area of triangle ace represents the real resource costs associated with increased domestic production.

Estimating Benefits: The Microeconomic Approach

Stockpile drawdowns provide benefits by reducing the costs of supply disruptions. The costs of supply disruptions can be estimated using the same approach employed to measure social surplus costs of tight market purchases. As we will subsequently discuss, the large price increases associated with supply disruptions raise some doubts about the appropriateness of the social surplus approach that are easily dismissed for the small price changes caused by tight market purchases. For very large price changes, the welfare interpretations of the standard supply and demand analysis become strained, and the consideration of macroeconomic effects becomes important.

In figure 4 a supply disruption of size S increases price from P(O) to P(S). Consumption falls from q_{D1} to q_{D2}, production increases from q_{S1} to q_{S2}. The area of the trapezoid, abhg, represents the social surplus loss of the disruption. The total consumer surplus loss is actually given by the area of P(O)bhP(S). This area overstates the social surplus loss, however, because the portion of this area to the left of the supply curve, P(O)agP(S), represents wealth transfers from consumers to domestic producers. Now assume that the stockpile is drawn down by an amount, A_2, so that the world price only rises to P(S-A_2) instead of P(S). Whereas the social surplus loss is the area of abhg without the drawdown, it is the area of abji with it. The difference between these two areas, ijhg, is the social surplus loss avoided by the drawdown.

To what extent are social surplus changes calculated in this way good measures of changes in social welfare? Our answer depends in several ways on the size of the price increases being analyzed.

To begin, consider how we would measure changes in the welfare of individual consumers in the general equilibrium context. When the price of one good is exogenously increased, consumers and producers will alter their behavior so that a new set of equilibrium prices will result. A consumer with a fixed income or budget will derive some level of welfare from the goods purchased at the original set of prices and some new level of welfare from the goods purchased at the new set of prices. A measure of the change in welfare experienced by the consumer is the amount of income that would have to be given to the consumer after the price changes to raise his welfare back to the original level. This amount is called compensating variation.[9] We could imagine adding the compensating variations for all consumers to obtain the change in aggregate (social) welfare caused by the exogenous price increase.

In practice we never have sufficient information to employ the general equilibrium approach. Instead, we usually assume we can use the partial equilibrium approach, limiting our attention to one market or perhaps a small number of markets. Assuming income and the prices of all other goods to be constant, we can often estimate a Marshallian (or constant income) demand curve for a good from empirical data. The demand curve shown in figure 4 is an example; we at least know how much is demanded at the current price and can use data from past price changes to guide us in our assumptions about the shape of the rest of the curve. If incomes of all consumers and the prices of other goods actually remained constant after the price of the good increased, the consumer surplus change measured using the Marshallian demand curve would be a close approximation of the sum of compensating variations. The smaller the price change considered, the closer the correspondence.[10]

In our problem oil is not a final consumption good. Rather, it is an input into the production of a number of final consumption goods such as gasoline, electricity, heating oil, and plastics. The demand curve for oil is derived from the demands for these final products. The direct partial equilibrium procedure would be to measure consumer surplus changes in each of the markets for these final consumption goods. It can be demonstrated, however, that if income effects are ignored, the change in consumer surplus measured in the input market will closely approximate the sum of the consumer surplus changes in the markets for the final consumption goods.[11] So for small price increases, it is reasonable to look at the oil market alone.

Large price increases in the oil market bring into question the appropriateness of the partial equilibrium approach. During a large supply disruption, prices and incomes will change throughout the economy. The relative prices of oil substitutes will increase, while the relative prices of goods complimentary to oil products will decline. The aggregate level of real income and its distribution will readjust as profits and payments to owners of factors of production change, causing shifts in the Marshallian demand curves we employ in partial equilibrium analysis. The larger the price increases, the more tenuous the link between social welfare losses and the social surplus changes we measure using the demand curve for oil. In light of macroeconomic effects of price shocks, the social surplus approach most probably underestimates the social welfare costs.

Estimating Benefits: The Macroeconomic Approach

As long as we import oil, a rise in the world price will reduce the aggregate supply of goods that can be produced at the prevailing general price level, assuming no change is made in monetary and fiscal policies. Because they must pay more for the factors of production, firms will not maintain their levels of output unless there is a rise in the general price level to cover their increased costs. Figure 5 illustrates this supply-side effect in the context of aggregate supply and demand, the simplest macroeconomic framework.[12] Before the oil price shock, the levels of real output, Y, elicited by various general price levels, P, are given by the aggregate supply curve labeled AS_0. The shock causes the aggregate supply curve to shift to the left to AS_1. If the aggregate demand curve, which represents the levels of real output demanded at various general levels, does not shift, the equilibrium output level will fall from Y_0 to Y_1, and the equilibrium general price level will rise from P_0 to P_1. A higher rate of unemployment will accompany the decline in real output.

But the oil price shock will also shift the aggregate demand curve leftward so that at any particular general price level, less real output will be demanded. The increased transfer of funds to foreign producers (area klhg in figure 4) directly reduces the real income of U.S. consumers and hence reduces their demand for all normal goods. The large increase in the funds flowing to oil producers reduces aggregate demand through what has been called the "oil price drag." Foreign producers will not immediately spend their increased revenues on imported goods or investments in the consuming nations. Consequently, there will be a delay between the time funds leave the United

States and the time they are recycled to the U.S. economy in the form of orders for goods and services or purchases of U.S. assets. Eventually the international monetary system will accommodate faster recycling, but in the meantime aggregate demand in the United States will fall. A similar effect will occur if domestic oil producers do not immediately use their increased revenues for purchases of goods and services. The large wealth transfers also may reduce demand for investment goods by creating uncertainty about the composition of consumer demand in the future. In addition to shifting the aggregate demand curve in the current period, lower investment reduces the productive capacity of the economy in the future.

The shift in aggregate demand is represented by AD_1 in figure 5. The new equilibrium occurs at the intersection of AD_1 and AS_1 with the general price level at P_2 and the real output level at Y_2. The shift in aggregate demand results in a lower level of real output and a lower general price level than would occur if only the aggregate supply curve had shifted. The actual location of the aggregate demand curve will depend on the monetary and fiscal policies adopted in response to the price shock. By shifting AD_1 toward AD_0, monetary and fiscal policies (such as expansion of the money supply or tax reductions) reduce the loss in real output at the expense of a higher price level.[13]

Figure 5
Oil Price Shocks in the Aggregate Supply and Demand Framework

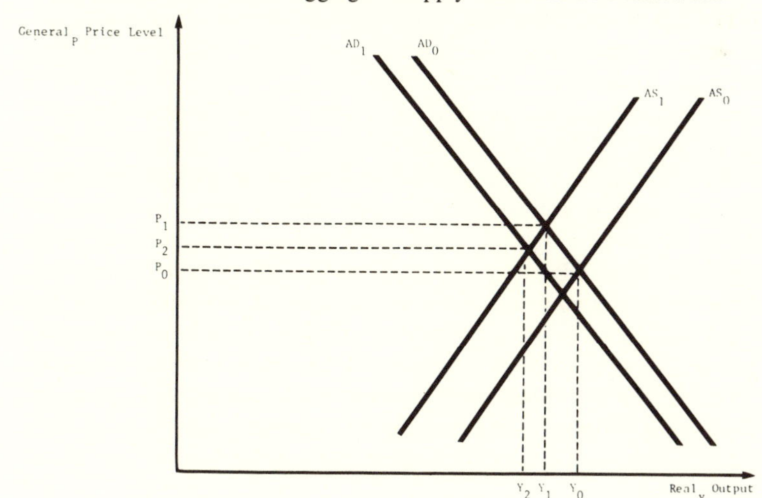

Stockpile drawdowns reduce the losses in real output and the increases in the general price level by limiting the size of the wealth transfers to foreign and domestic producers through reductions in the world price. In terms of our aggregate analysis, drawdowns prevent the aggregate supply curve from shifting all the way to AS_1 and the aggregate demand curve from shifting all the way to AD_1. The resulting equilibrium will be at a level of real output that is higher and a general price level that is lower than would occur in the absence of the drawdown. Noting that increased rates of unemployment will be associated with reductions in real output and giving a dynamic interpretation to our analysis, we recognize that the stockpile drawdown provides benefits in the form of lower levels of unemployment and inflation and increased rates of real economic growth.

Evaluation of various stockpile programs in the cost/benefit framework requires the measurement of benefits in dollars that are comparable to the dollars used to measure costs. Unfortunately, there is no obvious way to value the benefits of reductions in unemployment and inflation rates in dollars. Gross national product (GNP), adjusted for changes in the general price level, provides a dollar value of real output.[14] Unlike consumer surplus, however, no direct interpretation of changes in GNP as changes in social welfare can be made. Although larger social welfare will be associated with larger GNP losses, the relationship is not linear. Larger GNP losses represent more than proportionally greater social welfare losses.[15] Despite this conceptual difficulty, many of the cost/benefit analyses of the SPR have measured benefits as losses in real GNP avoided through drawdowns.

Large-scale macroeconomic models have been the primary source of estimates of the GNP benefits of stockpile drawdowns.[16] The models consist of hundreds of econometrically estimated equations that are used to simulate the future performance of the economy under a variety of assumptions about energy prices, monetary and fiscal policies, and stockpile drawdowns. To predict the benefits of a drawdown, the model is used to predict a GNP path under the hypothesized disruption with and without drawdowns. For the period of supply disruption and a period immediately following, the differences in real GNP are calculated. The benefit of the drawdown is the discounted value of the stream of differences.

There are three general problems associated with this methodology. First, the econometrically estimated relationships in the models may not hold for large disruptions beyond the range of historical data. Second, a large number of arbitrary adjustments of the models must be made to obtain internally consistent results for large disruptions. Third, the models focus primarily on aggregate demands with only implicit representations of supply flows and therefore may not provide accurate predictions for large supply-side shocks. The first two problems are unavoidable; the third can be mitigated somewhat by linking the models to an explicit input/output model of production.[17] In fact, some of the early stockpiling studies employed reduced form GNP loss functions derived solely within the input/output framework.[18]

In summary, there is no fully satisfactory way to measure the benefits of stockpile drawdowns during supply disruptions. Problems in measurement swamp the conceptual link between social surplus and social welfare for a large disruption; the conceptual connection between real GNP and social welfare is unclear. Despite these limitations, both approaches have been employed in a variety of ways.

Dealing with Uncertainty

Over the next fifteen to twenty years, the Middle East will continue to account for a large fraction of world oil production. During this period the Middle East will continue to be politically volatile. If these commonly advanced assertions are true, we face the possibility of experiencing major disruptions in the world oil market before the end of the century. The variety of circumstances that could lead to reductions in oil exports from the Middle East suggests that the probability of at least one major disruption sometime during this period is quite high. We cannot know with certainty how many, if any, disruptions will actually occur, when they will occur, how large they will be, or how long they will last. By making assumptions about the probabilities that various types of disruptions will occur, we can convert the problem from one of uncertainty to one of risk. Using the assumed probabilities, we can evaluate stockpiling programs in terms of expected costs and expected benefits.

The calculation of expected net benefit is conceptually simple if measures of costs and benefits are available. First, an exhaustive set of mutually exclusive future states of the world are specified. Second, the present value of the net benefits associated with each of the possible future states is determined. Third, the total present value of expected net benefits is calculated as the sum of the present value of expected net benefits associated with each state of the world weighted by the probability that the state will occur.

In practice, the enumeration of the analytically distinct states of the world and their probabilities is tedious and complex. Instead of attempting to identify a large number of the most probable states, most studies have employed what we term here the scenario/break-even methodology. They have focused on a single scenario in which a disruption of a certain size and duration is assumed to occur in some year in the future after the completion of the acquisition of the stockpile. The present value of the benefits from the drawdown of the stockpile during the disruption, B, is calculated. If C is the present value of the program costs, the present value of the expected net benefits of the stockpile program would be given by PxB-C where P is the probability that the scenario will occur, and (1-P) is the probability that no disruption will occur over the time frame of the program. As long as B is greater than C, the present value of expected net benefits will be greater than zero if P is greater than $P_{BE}=C/B$, the break-even probability. Because the scenario/break-even methodology focuses on a single scenario, it makes practical the use of the large-scale macroeconomic models for measuring benefits. It also draws attention to the assumed probability of disruption as a critical factor in the determination of the desirability of the program.

The scenario/break-even methodology oversimplifies the stockpiling analysis problem in a number of important ways. It does not allow for uncertainty over the timing, size, and duration of disruptions. It ignores the possibility of there being a disruption before the acquisition of stocks has been completed. It ignores the possibility of there being more than one disruption over the life of the program. In evaluating the results of scenario/break-even analyses, decision-makers must subjectively account for these limitations in deciding how large a probability to assign to the occurrence of the scenario. Additionally, the scenario/break-even methodology provides little insight into strategies for achieving the optimal timing of acquisitions and drawdowns.

Decision analysis provides a framework for more explicitly taking account of the various aspects of risk involved in the stockpiling problem. The life of the program is divided into a number of stages. At each stage a decision is made concerning changes in the size of the stockpile, given the current state of the market. The decisions translate into costs and benefits, depending on the state of the market that randomly occurs in the next stage. This decision process can be displayed in a decision tree, which provides a diagrammatic map for systematically calculating the expected net benefits of each possible decision at each stage.

A simple three-stage decision tree is displayed in figure 6. The state of the world at the beginning of each stage is described by the size of the existing stockpile, S, and the state of the market, M. The possible stockpile sizes are zero, one, or two units. The market can be either disrupted (M=D) or nondisrupted (M=N). The state of the world at the beginning of each stage is given by a pair (S,M). At the beginning of the first stage, for example, the state of the world is (O,N). If the market is nondisrupted, it is assumed that either a unit of stocks is purchased (A=1) or not purchased (A=0). If the market is disrupted, it is assumed that either a unit of stocks is sold (A=-1) or not sold (A=0). In the third stage it is assumed that all stocks are sold no matter what the state of the market because the stockpile is assumed to be no longer needed (and for expository convenience).

At each stage it is assumed that there is a probability, Pr(M=D)=P, that the market will be disrupted. The state of the world in the next stage depends on the stock change decision made at this stage and the random occurrence of the next market state. For example, starting in state (O,N), purchase of a unit of stocks, A_1(O,N)=1, has a probability P of resulting in state (1,D) and a probability 1-P of resulting in state (1,N) in the second stage.

Figure 7 illustrates the use of the decision tree for calculating expected net benefits. The probability of a disrupted market is assumed to be 0.1. The costs of the possible decisions concerning changes in the size of the stockpile for each of the market states are displayed in the table at the bottom of figure 7. For example, as calculated from the table of net costs, a disruption inflicts costs of 100 without a drawdown and costs of 60 with a drawdown of one unit,

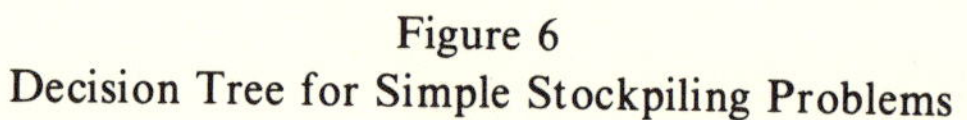

Figure 6
Decision Tree for Simple Stockpiling Problems

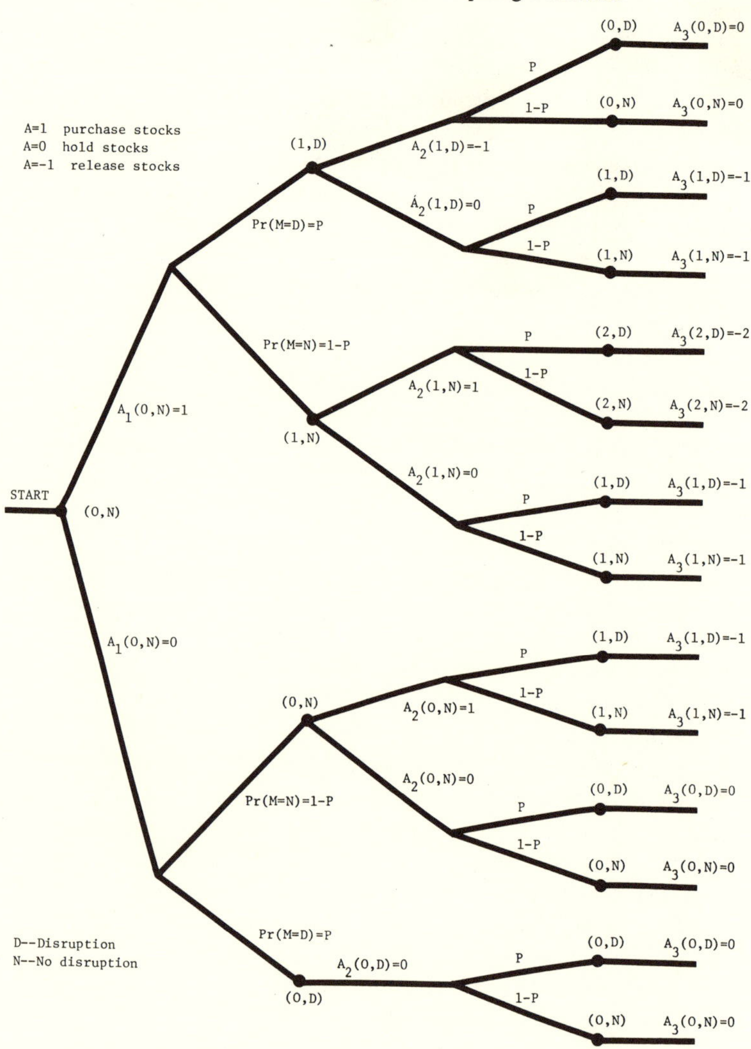

Figure 7
Numerical Example of Decision Tree Analysis

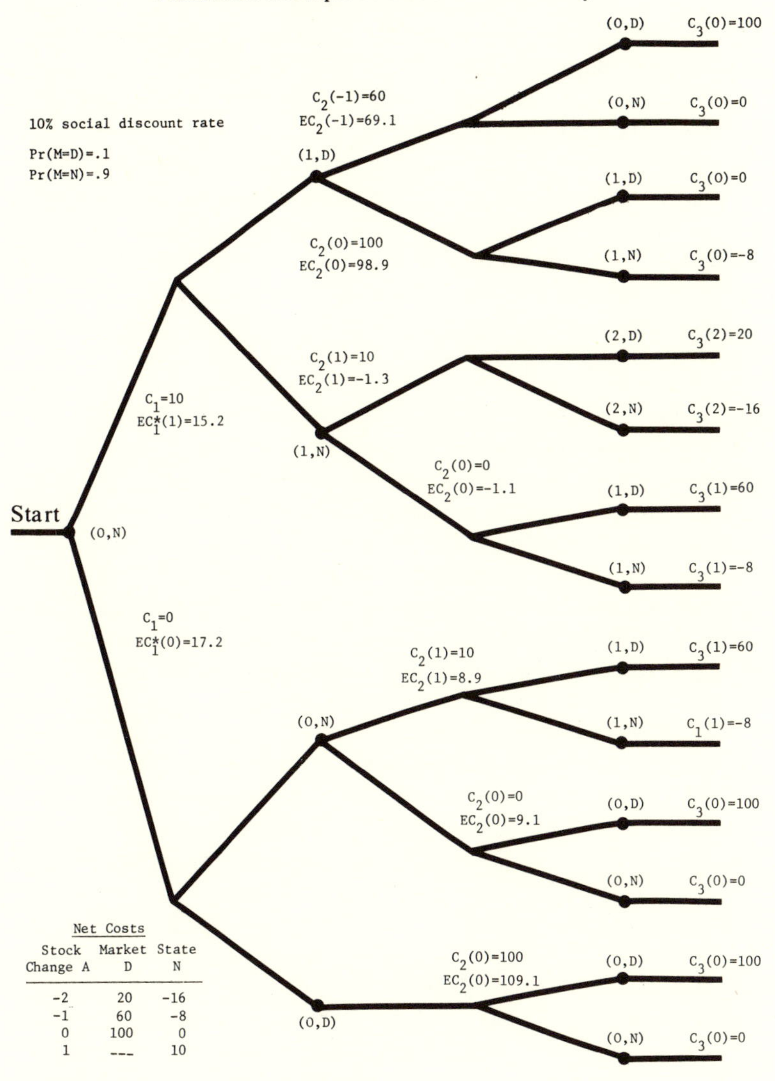

D-Disruption N-No Disruption

implying net benefits for the drawdown of 40. The immediate cost of a decision A at stage i is given by $C_i(A)$. The present value of expected net costs of a decision A at stage i is given by $EC_i(A)$, which is the sum of the current costs and the present value of expected future costs. For example, $EC_2(-1)$ for state (1,D) is 60 + [(0.1) (100) + (0.9) (O)] / 1.1=69.1. The expected value of net costs for the alternative choice is EC (0)=98.9. Because 69.1 is smaller than 98.9, the optimal decision at this stage given state (1,D) is to draw down the stockpile. In general, the optimal decision at each stage is contingent upon the current state of the market and the current stockpile size.

We can find the optimal initial decision assuming that optimal decisions are made at subsequent stages. We label the present value of expected net costs of the decision at the first stage as EC * (A) to signify that it is assumed that optimal decisions will be made in the following stages. In our example, the present value of expected net costs of purchasing in the first stage is 15.2, whereas the present value of expected net costs of not purchasing in the first stage is 17.2. The analysis also indicates that the largest optimal size of the stockpile is 2 units. It would be optimal to have a stockpile of this size if a purchase was made in the first stage and no disruption occurred so that the optimal decision in the second stage was also to purchase. The decision analysis methodology thus indicates the largest optimal stockpile size as well as the optimal acquisition and drawdown decisions at each stage for each state of the world.

Our stylized example was sufficiently small to allow us to draw the decision tree and calculate the present values of expected net costs at each stage. For a more realistic model of the stockpile problem, we might have twenty or more stages (one for each year of the program's life), four or more market states (slack market, tight market, minor disruption, major disruption), and forty or more stockpile sizes (say 50 million barrel increments up to 2 billion barrels). It is not practical to solve such a problem by calculating present values of expected net costs for each possible decision in the decision tree. Fortunately, there is a method called dynamic programming that can be employed to solve large decision-tree problems with the use of a computer.[19]

The decision analysis/dynamic programming methodology was first applied to the strategic stockpiling problem by Thomas J. Teisberg of Massachusetts Institute of Technology. Working with Glen Sweetnam, Steven Minihan, and George Horwich of the Office of

Policy and Evaluation in DOE, he developed a model that measures benefits and costs in terms of social surplus changes. The model employs assumptions about the probability of moving from each market state to each of the other market states. It calculates the optimal acquisition/drawdown decision in each period as a function of the state of the market and the entering stockpile size. Recently other researchers have employed the dynamic programming methodology measuring costs with a macroeconomic loss function.[20]

Summary of Basic Assumptions

The numerous cost/benefit evaluations of the SPR can be classified along two dimensions: the measurement of benefits and the handling of uncertainty. Within this classification, two approaches have been prominent in the policy debate. Scenario/breakdown studies have been repeatedly employed. More recently, the decision analysis/dynamic programming methodology employing social surplus measures of benefits has been advanced as an alternative.

Even after agreement has been reached on the methodological approach to be employed, there is room for considerable controversy over the numerous assumptions that must be made. Table 4 reviews the major assumptions we have discussed. It also indicates how the assumptions influence the results of the cost/benefit analyses of variously sized stockpiles.

Analysis in the Bureaucratic Environment

In the next chapter we consider the role of economic analysis in the resolution of the SPR size issue. Our story is one of bureaucratic politics where analysis serves as the focus of conflict. Hence, much of the presentation is from the point of view of the analyst. Unfortunately, such a story cannot be told without considerable detail, which may distract attention from the more generic issues involved. We offer here some general observations on the nature of analysis in the bureaucratic setting to help place the substantive story in a more general context.

The first observation is obvious but nevertheless important: no quantitative economic analysis is ever perfect. Even the most sophisticated analyses must reduce the complexity of the real world through

Table 4
Review of Major Analytical Assumptions Made in SPR Evaluations

ASSUMPTION	*COMMENT*	*SIGNIFICANCE FOR STOCKPILE SIZE (CETERIS PARIBUS)*
1. GENERAL		
underlying oil price trajectory	backstop and rate of growth in real prices	steeper trajectories favor large stockpiles because of higher scrap value of oil at program termination
short-run world demand curve for oil in each period	determination of world consumption in normal markets and world price in disrupted markets	more inelastic curves favor larger stockpiles
world supply response to stockpile purchases	definition of slack and tight markets	slacker markets favor larger stockpiles and their use in smaller disruptions
price ratchet after disruption	usually no ratchet assumed; nominal price ratchet may be reasonable	ratchets favor larger stockpiles by extending duration of direct benefits of drawdowns
storage facility costs	timing and costs of storage capacity development	lower costs favor larger stockpiles, earlier availability may do so, also
social discount rate	calculation of present value of future costs and benefits	lower social discount rates favor larger stockpiles because costs tend to precede benefits
disruption scenarios	sizes, durations and frequencies of representative set of disruptions; use of stockpiles by U.S. and other consuming nations	larger, longer, and more frequent disruptions favor larger stockpiles
2. MICROECONOMIC APPROACH		
short run U.S. demand curve for oil in each period	measurement of social surplus; determination of U.S. share of shortfall	more inelastic curves favor larger stockpiles

Table 4 *continued*

ASSUMPTION	*COMMENT*	*SIGNIFICANCE FOR STOCKPILE SIZE (CETERIS PARIBUS)*
short run U.S. supply curve for oil in each period	measurement of social surplus and size of U.S. shortfall	more inelastic curves favor larger stockpiles
3. LARGE SCALE MACROECONOMIC MODELS		
monetary and fiscal policy responses to disruptions	determination of combination of inflation and real GNP losses resulting from disruption	expansionary monetary and fiscal policies will favor smaller stockpiles because inflation is not explicitly in loss function
exogenous factors	prices of other energy inputs, demand for exports, changes in propensity to save	higher prices for energy substitutes favor larger stockpiles
adjustments to correct for supply side weaknesses	corrections so oil consumption does not exceed supply during disruptions; modification of variables not consistent with modelers' preconceptions	depends on particular adjustments
4. REDUCED FORM GNP MODELS		
GNP loss function	GNP as function of world price or U.S. shortfall	more severe loss functions favor larger stockpiles
5. DYNAMIC PROGRAMMING		
matrix of transition probabilities among market states	given the current market state, the probabilities of being in each of the possible market states in the next period; deterrence can be modeled by making probabilities function of stockpile size	higher probabilities for disruptions and slack markets favor larger stockpiles

numerous approximations and assumptions. When one faction advances analysis in support of some policy action, factions in opposition can always point to some analytical weakness as a rationale for delay. In fact, a strategy for countering a political predisposition for a particular action is to force proponents to back their position with analysis. The analysis will undoubtedly have weaknesses that can be exploited in an attempt to delay or defeat the proposed policy action.

The second observation is related to the first: political decision makers are interested in analytical results; they rarely have the inclination, time, or expertise to give attention to the technical aspects of analysis. Consequently, the analysts themselves usually must resolve disputes over approaches and assumptions. Without a consensus among the analysts over the appropriate resolution of technical issues, the results are unlikely to have much influence on decision makers. When analysts bring conflicting policy preferences to their work, it is difficult to reach a consensus. This is especially true when the analysis is being conducted within the bureaucracy and away from the view of professional peers, who might hold the participants to certain standards.

Third, the bureaucratic environment discourages the introduction of new methodologies. The greater the consensus among opposing factions over the appropriateness of basic approaches and assumptions, the greater the likelihood that the analysis will influence decision makers. Proponents of actions will favor methodologies that opponents have previously accepted in order to limit the range of controversy that must be resolved. Participants who view the analytical process primarily as an unavoidable ritual are unlikely to be strongly motivated to invest in the development and introduction of methodologies solely because they have potential for providing a better answer, especially after effort has been invested in learning how to perform (and manipulate) a methodology that already has been accepted. Unlike academic researchers, analysts in the bureaucracy are not necessarily rewarded for bringing fresh approaches to old problems.

Finally, political constraints, such as enabling legislation and budget limitations, tend to restrict the range of alternatives that are considered. The general approach of economics is to find the alternative that is socially optimal. The approach of analysts in the bureaucracy is more often to select the best alternative among those perceived to be politically feasible. Also, questions of optimal program size and configuration are never resolved. Rather, analysis becomes the focus of conflict in successive budget cycles.

Notes

1. Egan Balas has developed a game theoretic model for analyzing potential conflict between the U.S. and an oil cartel. He assumes the U.S. minimizes the sum of costs of the embargo and the SPR. He also assumes the cartel attempts to maximize the costs inflicted on the U.S. less their own costs. His results for particular scenarios include the loss-absorption and embargo-deterrence effects of the SPR. The approach is limited because it does not incorporate nonembargo disruption scenarios and because it employs an arbitrary objective function for the cartel. It is therefore of conceptual interest but not readily incorporated in the general cost/benefit framework. Egan Balas, "The Strategic Petroleum Reserve: How Large Should It Be?" Management Science Research Report No. 436, Graduate School of Industrial Administration, Carnegie-Mellon University, June 30, 1979.

2. Analyses using the dynamic programming model developed by Thomas J. Tiesberg have looked at private sector displacement and coordinated stockpiling. Thomas J. Teisberg. "A Dynamic Programming Model of the U.S. Strategic Petroleum Reserve," *Bell Journal of Economics* 12, no. 2 (Autumn 1981): 526-46. For an example of the international cooperation question see William W. Hogan, "Oil Stockpiling: Help Thy Neighbor," Energy and Environment Policy Center, John F. Kennedy School of Government, Harvard University, Cambridge, Mass., March 1982.

3. Harold Hotelling, "The Economics of Exhaustible Resources," *The Journal of Political Economy* 39, no. 2 (April 1931): 137–75. For a general review, see Frederick M. Peterson and Anthony C. Fisher, "The Exploitation of Extractive Resources: A Survey," *Economic Journal* 87 (December 1977): 681–721; and Shantayanen Devarajan and Anthony C. Fisher, "Hotelling's 'Economics of Exhaustible Resources': Fifty Years Later," *Journal of Economic Literature* 19, no. 1 (March 1981):65-73.

4. Where extraction costs are a function of cumulative production, the marginal costs of extracting a unit of resource in a particular period will include a term representing the increased costs of extraction in all future periods. See for example, Geoffrey Heal, "The Relationship Between Price and Extraction Cost for a Resource With a Backstop Technology," *Bell Journal of Economics* 7, no. 2 (Autumn 1976): 371-78.

5. More disconcerting for the applicability of the pure theory of exhaustible resources is the apparent fall in real prices over time of almost all

exhaustible resources. See Julian L. Simon, "Global Confusion 1980: A Hard Look at the Global 2000 Report," *Public Interest*, no. 62 (Winter 1981), pp. 3-20.

6. See Walter J. Mead, "An Economic Analysis of Crude Oil Price Behavior in the 1970's," *Journal of Energy and Development* 4, no. 2 (Spring 1979):212–28; and Ali D. Johany, "OPEC and the Price of Oil: Cartelization or Alteration of Property Rights," *Journal of Energy Development* 5, no. 1 (Autumn 1979):72–80.

7. If the supply is fixed at S, the price will be given by $P=D(S)$ where D is the inverse demand function. If an acquisition, A, is made in the absence of a supply response, the new price will be $P'=D(S-A)$. The greater elasticity of demand, the smaller the increase in price, $P-P'$.

8. It has been argued that the marginal social cost of oil to the U.S. is above the world price by a premium that reflects the impact of the marginal barrel of imported oil on the world price (monopsony premium) and the reduced vulnerability of the U.S. to disruptions from lower import levels (disruption premium). The disruption premium is explicitly accounted for in the cost/benefit approach employed. In slack markets, there is no monopsony premium because price is invariant to the marginal purchases. In tight markets, there is a price effect that must be explicitly included. For a general review of the premium issue, see William W. Hogan, "Import Management and Oil Emergencies," in *Energy and Security*, ed. David A. Deese and Joseph S. Nye (Cambridge, Mass.: Ballinger Publishing Company, 1981), pp. 261–301.

9. An alternative exact measure of welfare change is equivalent variation, the amount of income that would have to be taken away from the consumer facing the old prices to make him indifferent between the old prices with the adjusted income and the new prices with the original income. See Hal R. Varian, *Microeconomic Analysis* (New York: W. W. Norton & Company, 1978), pp. 208–11. For discussion of the use of consumer surplus in cost/benefit analysis see Arnold C. Harberger, "Three Basic Postulates of Welfare Economics: An Interpretative Essay," *Journal of Economic Literature* 9, no. 3 (September 1971):785–97.

10. See Robert D. Willig, "Consumer's Surplus Without Apology," *American Economic Review* 65, no. 4 (September 1976):589–97.

11. Tani and Boyd demonstrate this result in the context of a general equilibrium model where the consumer surpluses in the final demand markets are measured under the curve that connects the old and new equilibria. As such, the consumer surpluses do not correspond to those we would measure in a partial equilibrium analysis. Therefore their analysis should be taken as an indication of the approximate correspondence between the consumer surplus measured in the input markets and those measured in the markets for the final goods. Steven N. Tani and Dean W. Boyd, "Measuring

the Economic Cost of an Oil Embargo," Stanford Research Institute, Stanford, Calif., October 1976, pp. 28–38.

12. For an expanded discussion of oil price shocks in the aggregate supply and demand framework, see George Horwich, "Government Contingency Planning for Petroleum Supply Interruptions: A Macro Perspective," presented at the Conference on Policies for Coping with Oil Supply Disruptions, American Enterprise Institute for Public Policy Research, Washington, D.C., September 8-9, 1980; Edward M. Gramlich, "Macro Policy Responses to Price Shocks," in *Brookings Papers on Economic Activity*, ed. Arthur M. Okun and George L. Perry, no. 1 (1979), pp. 125–66; Knut Mork and Robert Hall, "Macroeconomic Analysis of Energy Price Shocks and Offsetting Policies: An Integrated Approach," in *Energy Prices, Inflation, and Economic Activity*, ed. K. A. Mork (Cambridge, Mass.: Ballinger Publishing Company, 1981), pp. 43–62; William D. Nordhaus, "The Energy Crisis and Macroeconomic Policy," *The Energy Journal* 1, no. 1 (January 1980):11–20; and Robert S. Pindyck, "Energy Price Increases and Macroeconomic Policy," *The Energy Journal* 1, no. 4 (October 1980):1–20.

13. Tax cuts that reduce the payroll or other business taxes will also shift the aggregate supply curve. Reductions in payroll taxes also have the advantage of counteracting rigidities in wages that contribute to unemployment. See Mork and Hall, "Macroeconomic Analysis of Energy Price Shocks and Offsetting Policies."

14. Gross national product is measured according to the accounting identity GNP=C+I+G+(X-M) where C is consumer expenditure, I is gross business investment, G is purchases of goods and services by government, X is exports, and M is imports.

15. GNP is a measure of the level of economic activity. It is the sum of the market values of all the goods and services produced in the economy. But consumers place a somewhat higher value on the goods and services than their market value or they would not purchase them. The first marginal reduction in the goods and services traded takes them away from consumers who valued them at close to their market prices. The next marginal reduction takes goods and services away from consumers who placed a somewhat higher value on them. Greater compensation would have to be given to consumers to restore them to their levels of welfare after the second marginal reduction than after the first.

16. Most commonly used are the major commercial forecasting models operated by Data Resources, Inc. (DRI), Wharton Econometric Forecasting Associates, and Chase Econometrics.

17. For discussion of these efforts see Edward R. Novicky, "A Review of Analytical Techniques and Studies Related to Assessing the Impacts of Petroleum Shortfalls," STSC, Inc., Management Technology Division, Bethesda, Maryland, May 1979, pp. 67–84.

18. Of greatest practical relevance is the early work of Randall G. Holcombe which was based on data from the 1973 Arab embargo. Randall G. Holcombe, "A Method for Estimating the GNP Loss from a Future Oil Embargo," *Policy Sciences* 8, no. 1 (June 1977):217–34.

19. For an introduction to decision analysis, see Howard Raiffa, *Decision Analysis* (Reading, Mass.: Addison-Wesley, 1968). For an introduction to dynamic programming, see Frederick S. Hillier and Gerald J. Lieberman, *Introduction to Operations Research* (San Francisco: Holden-Day, Inc., 1972), pp. 239–64.

20. Hung-po Choa and Alan S. Manne, "Oil Stockpiles and Import Reductions: A Dynamic Programming Approach," Electric Power Institute, Palo Alto, California, October 1980; and James L. Plummer, "Methods for Measuring the Oil Import Reduction Premium and the Oil Stockpile Premium," *The Energy Journal* 2, no. 1 (January 1981):1–18.

6

The Size Issue and OMB: Analysis in the Bureaucracy

To the casual observer of national energy policy, it might appear that the issue of the ultimate size of the strategic petroleum reserve has been settled. In April 1977 the Carter administration issued the first National Energy Plan (NEP I), which called for the establishment of a one billion barrel reserve by the end of 1985.[1] Congress endorsed this goal by allowing Amendment Number 2 to the SPR Plan to become effective on June 13, 1978.[2] The one billion barrel goal was reaffirmed as administration policy a year later in NEP II.[3] The size issue would thus seem to have been resolved at the highest government levels. But a goal is virtually meaningless without a commitment of the resources needed to achieve it. The billion barrel SPR cannot be realized unless Congress authorizes and appropriates funds for the construction of storage facilities. Generally critical of the slow pace of the SPR program, Congress has, nevertheless, shown a willingness to meet administration requests for funding. It was the Carter administration itself that showed hesitancy in seeking funds.

Within the administration there was continuous conflict over the goal. On one side was DOE, pursuing it despite the problems encountered in implementing the first 500 million barrels of storage (Phases I and II); on the other was OMB, trying to overturn or delay its realization by forcing reconsideration of the size issue in repeated budget cycles. OMB was instrumental in decisions to delay by two years implementation of the third 250 million barrels of capacity (Phase III) and to postpone indefinitely implementation of the fourth 250 million barrels of capacity (Phase IV). How should we view the OMB challenge of the presidentially established and congressionally endorsed SPR goal?

OMB is part of the Executive Office of the president.[4] Its most visible role is to assist the president in preparing and administering the budget. More generally, it provides the president with assessments of executive branch operations, including evaluations of specific programs and policies. OMB attention to the SPR program, both in terms of management and size issues, was appropriate for two primary reasons.

First, early cost overruns and delays brought into question the validity of DOE estimates of the costs of and feasible schedule for future SPR expansion. Higher costs and delayed availability of storage capacity could alter estimates of the net benefits of the billion barrel SPR. Programmatic problems also justified reconsideration of the industrial petroleum reserve option and other alternatives. For example, perhaps a program to induce or coerce private sector stockpiling might appear more attractive in light of implementation problems encountered by DOE.

Second, changing perceptions of the future energy picture suggested the need to reanalyze the size issue. Higher oil prices could be expected to lead to lower levels of imports in the future; the president's energy program might do so also. The Camp David Accords might reduce the chances of market disruptions related to conflicts between Israel and its neighbors. These and other observations of the then current scene were advanced as rationales for reopening the size issue.

Other factors help explain OMB efforts to restrict the size of the SPR. The development of storage facilities and the purchase of oil involve large budgetary expenditures. The benefits of these immediate expenditures, although large in an expected value sense, may or may not actually be realized in the future. Increasing pressure to balance the budget as the 1980 election approached brought into question the relative political value of expanding the SPR. Because its benefits are diffuse, the SPR has not acquired a constituency willing to expend resources in fights over budget cuts. Consequently, it is a more inviting target for budget cutters than other programs enjoying the vocal support of organized beneficiaries.

Personal motivations may have been an important factor in the persistence of OMB in pursuing the size issue. As the White House energy adviser, James Schlesinger outflanked OMB by securing approval for the billion barrel goal directly from the president during preparation of NEP I. OMB staffers were later cut short by the president when they attempted to reverse the goal during planning for the 1979 fiscal year budget. A desire to avenge these bureaucratic defeats may have encouraged their demands for repetition of the size studies.

A large number of analysts, both inside and outside the government, have contributed to the debate over the appropriate size of the SPR.[5] Our focus in the following discussion will be primarily the analyses conducted within the administration, with emphasis on those that were particularly important in the debate between DOE and OMB. After reviewing the studies from the analysts' perspective, we will attempt to place them in the context of the larger policy process.

Battlelines are Drawn

EPCA declared the policy of the United States to be creation of petroleum reserves of up to one billion barrels. It presumed that the SPR Plan would provide for the implementation of a reserve corresponding to about ninety days of U.S. imports (approximately 500 million barrels) within seven years. Although the selection of these size parameters was not the result of explicit cost/benefit analyses, two studies were influential.

Beginning in 1972, the National Petroleum Council, then the industry advisory group to the Department of the Interior, investigated U.S. vulnerability to disruptions of petroleum imports. A postmortem of the Arab embargo was completed in September 1974.[6] It recommended the development of salt dome storage capacity for emergency reserves of 500 million barrels. The recommendation, based on the assumption that drawdowns of emergency reserves and stocks held by industry would replace import shortfalls barrel for barrel, was intended to provide full protection against disruptions of up to three million barrels per day lasting up to six months. No allowance was made for the leakage of a fraction of the drawdowns out of the United States through the displacement of imports. That is, it was assumed that each barrel of oil released from the reserve would reduce the U.S. shortfall by one barrel. Although estimates of the costs of shortfalls to the economy and the costs of developing the emergency reserves were provided, no attempt was made to justify the recommendation as cost-effective. Nevertheless, the prestige of the National Petroleum Council gave legitimacy to the idea of developing reserves on the order of 500 million barrels.

The first attempt to find the cost-effective reserve size was made as part of the Project Independence effort completed under President Ford. Reserve sizes that minimized the sum of reserve program costs and GNP losses for selected disruption scenarios were determined. Program costs were estimated for a ten-year program life. The addition of these costs to GNP losses assumed that each disruption scenario under consideration occurred only once, with the probability of one during the ten-year period. The scenarios assumed that the United States would lose a fraction of its imports from insecure sources. It was assumed that the first million barrels per day of imports lost could be accommodated without loss in GNP. Based on estimates reported in the National Petroleum Council study, it was assumed that each million barrels per day lost above the first million would result in a GNP loss of $33 billion annually. Drawdowns were assumed to replace shortfalls barrel for barrel. Under scenarios with total U.S. imports of six million barrels per day, the minimum cost stockpile ranged from 730 million to 1,460 million barrels.[7] For higher import levels, even larger stockpiles minimized the sum of costs.

An office was established in FEA to develop legislation for the creation of a strategic reserve and to begin studies to prepare for its implementation. The office was instrumental in securing the flexibility found in the SPR provisions of EPCA. The early storage of 150 million barrels was based on an estimate made by the office of the amount of existing salt dome storage capacity that could be obtained for the program. Robert L. Davies, who headed the office, remembers the one billion barrel upper limit as coming largely "out of thin air."[8] Although work completed at the Institute for Defense Analyses suggested a much larger SPR was justified, the billion barrels was a round number consistent with the Project Independence analysis.[9] There was disagreement between FEA and OMB over the presumed size of the SPR. The goal of an equivalent of ninety days of imports was the compromise size selected by President Ford that became incorporated in EPCA.

While EPCA was taking form in Congress, analysts in FEA continued to work on the size issue.[10] A cost/benefit model that measured benefits in terms of avoided consumer surplus losses was developed.[11] It was employed in investigations of stockpile benefits in combination with demand reduction measures. These investigations suggested that a reserve of only several hundred million barrels would be adequate if accompanied with strong emergency demand reduction measures.[12]

During preparation of the SPR Plan in 1976, the newly created SPR Office conducted analyses to determine if other than a 500 million barrel reserve was justified. Benefits were measured using a GNP-loss function originally developed from an input-output model of the economy built by Randall G. Holcombe for the Center for Naval Analysis.[13] GNP losses were calculated as a function of the size of the import shortfall and the demand for petroleum prior to the disruption. A large-scale energy model, the Project Independence Evaluation System (PIES), was used to generate high and low domestic supply and demand forecasts for 1980 and 1985. Shortfalls of from one to seven million barrels per day for 180 days were modeled. Annual costs for each of the fifteen program-years were estimated for the various reserve sizes considered.

An effort was made to take account of the uncertainty about when the hypothesized disruption would occur. Timing was analytically important because the reserve was assumed to be developed at a linear rate to its ultimate size; the entire reserve would therefore not be available for use in disruptions occurring in the early years of the program. For each scenario the present value of net benefits was calculated for different assumptions about the year the disruption being considered occurred. These estimates were then averaged to arrive at an overall present value of net benefits for the scenario.

The probable use of the SPR in light of the uncertainty over the duration of disruptions was reflected in the assumed drawdown strategy. The daily drawdown rate would be the smallest of the daily shortfall, 3.3 million barrels, and 1 percent of the preceding day's remaining reserves. Under this strategy, some oil would always be held in reserve as insurance against an extended disruption. After the disruption, the stockpile would be replenished as quickly as possible.

The results were presented in the SPR Plan in order to support the 500 million barrel reserve size.[14] A reserve of 500 million barrels was found to be cost-effective for disruptions as small as 1.3 million barrels per day, assuming low import levels. Higher import levels or larger disruptions were found to justify larger reserves. With the exception of the assumption that drawdowns would replace shortfalls barrel for barrel, the analysis and its interpretation seem to be conservative with respect to size. In terms of break-even probabilities, at low levels of imports, a 5.5 percent yearly probability of a six-month disruption of 4.4 million barrels per day would yield a present value of expected net benefits of zero for an 800 million barrel reserve. Even the yearly probability of 5.5 percent implies a more than 42 percent probability of there being no disruption over the 15-year program life.

Although the results could reasonably have been advanced in support of a larger reserve, the FEA administrator decided not to recommend a reserve larger than 500 million barrels.[15] OMB was opposed to a larger reserve. In fact, some of the budget analysts at OMB talked about limiting the size to 200 million barrels, but they could not gain support for their position because of the presumption in the legislation for 500 million barrels. OMB opposition would have forced FEA to take the issue to the president. Because the administration was still talking about energy independence as if it were a realistic middle-range goal, seeking a larger SPR based on the assumption of high import levels through the 1980s would have been awkward. The SPR Plan recommended, without OMB objection, a reserve size of 500 million barrels.

In the early days of the new administration, James R. Schlesinger served as President Carter's chief energy adviser. As a former secretary of defense, he emphasized the national security aspects of energy policy. Believing reliance on Middle Eastern oil to be a weak link in our defense posture, he favored a larger SPR, whether or not it could be shown to be cost-effective solely in terms of direct economic benefits.[16] During preparation of the National Energy Plan, the energy policy overview for the new administration, he persuaded President Carter to seek expansion of the SPR to one billion barrels.

The development of the National Energy Plan was tightly controlled by Schlesinger. His staff tapped the agencies and departments for information but remained secretive about the policy options being considered. Operating efficiency was selected over "the more time-consuming process of consensus building."[17] The National Energy Plan draft was circulated within the administration for comments only a short time before President Carter released the final version on April 29, 1977. Consequently, OMB career staffers concerned with the SPR did not have time to mobilize their new director to seek a reconsideration of the billion barrel goal by the president. In bureaucratic slang, Schelesinger had successfully "rolled" OMB on the SPR issue.

OMB counterattacked during preparation of the administration's FY 1979 budget proposals in the fall of 1977. Schlesinger, as secretary of the newly created Department of Energy, had requested funding for construction of the second 500 million barrels (Phases III and IV) of storage capacity. When OMB cut these funds, Schlesinger appealed to the president. Schlesinger presented his case to the president at a White House meeting with the OMB director and members of his staff.[18] The OMB staffers, armed with flipcharts, began to make the case that a reserve size larger than 500 million barrels was unjustified. They were cut short, however, by President Carter, who stated that the size of the SPR was not at issue; he was still committed to an ultimate reserve size of one billion barrels. The real issue was how much would be spent in FY 1979 for Phases III and IV. As a compromise, the president allowed planning funds for only Phase III to be included. The results of the meeting might be viewed as a pyrrhic victory for Schlesinger. He secured reaffirmation of the billion barrel goal. In the process, however, OMB staffers, who would be involved in future budget fights over the SPR, were embarrassed in front of the president. It would be surprising if the incident did not influence their attitudes toward the SPR program.

In April 1978 DOE transmitted to Congress an amendment to the SPR Plan calling for expansion of the reserve to one billion barrels by the end of 1985.[19] The amendment described the expansion as a "major national security measure" justified by more realistic estimates of import levels and consideration of more serious disruption scenarios. Although cost/benefit analysis was not presented in the amendment, revisions of the analysis presented in the SPR Plan were provided in congressional testimony.[20] The Senate rejected the amendment at the urging of Senator Kennedy and other New England senators, who objected to provisions calling for regional storage only if it was less expensive than the primary storage capacity for crude oil. The amendment was resubmitted without the stated restriction on reserves in May and approved by Congress on June 13, 1978.

The Macro-Model Quagmire

In February 1978 OMB organized an interagency task force on contingency planning. It consisted of personnel from the Office of Contingency Planning in the Policy and Evaluation Office of DOE, the Special Studies Division for Natural Resources, Energy, and Science in OMB, and the Council of Economic Advisors (CEA). DOE explicitly agreed to analyze alternatives for implementation of the expanded reserve, including financing options and a mandatory industrial reserve program.[21] By agreeing to develop mutually acceptable shortfall scenarios and to consider the potential for private sector inventory drawdowns during disruptions, DOE implicitly agreed to reconsider the size issue.

By mid-April tentative agreement was reached among the participants on the assumptions that would be used in efforts to measure the macroeconomic costs of oil supply shortfalls. It was decided that the macroeconomic modeling would focus on U.S. shortfalls of two million barrels per day lasting six months and twelve months, and four million barrels per day lasting six months. OMB employed the Chase Econometrics Long Term Model; CEA, the DRI Quarterly Macroeconomic Model; and DOE, the Holcombe GNP-loss function. At the same time, the SPR Office was funding a study of the macroeconomic effects of disruptions based on the DRI model and the Energy Information Administration of DOE was conducting a similar study using both the DRI and Wharton Econometric Forecasting Associates models. Each of these later studies concluded that a billion barrel or larger reserve was justified, assuming one disruption occurred during the study period.[22] Although general agreement was eventually reached on the GNP-loss function that would be used in cost/benefit analysis of the SPR, DOE and OMB could not agree on other methodological issues.[23]

In October OMB challenged the DOE justification for the fourth 250 million barrels of storage. Questions were raised about the scrap value of oil storage facilities (revenues from the sale of assets at program termination), the disruption scenarios considered, and the endurance, or inventory drawdown capabilities, of the private sector.[24] The assistant secretaries for Resource Applications and Policy and Evaluation responded to the questions but failed to satisfy OMB objections.[25] Finally, OMB presented its own analysis arguing against Phase IV in a memorandum to the President.

Secretary Schlesinger prepared to argue the issue with OMB during meetings on the FY1980 budget. Schlesinger had been informed by his staff that OMB had asserted that the probability of a severe disruption (approximately the loss of 60 percent of Persian Gulf exports for six months) would have to be between 10 percent and 25 percent per year over twelve years to justify Phase IV. With CEA concurrence, Schlesinger presented the DOE case that a yearly probability of only 1 percent would justify Phase IV.[26] When OMB agreed (its figures had been for the entire twelve-year period), Schlesinger was thrown off balance. Rather than threaten to take the issue to the president, he agreed not to press for Phase IV planning funds for FY1980 and to recommend that the issue of ultimate size be referred to the National Security Council (which eventually recommended reaffirmation of the billion barrel goal). One OMB analyst who attended the meeting believes that OMB would have backed down if Schlesinger had demanded a hearing before the president where he would ask the secretaries of Defense and State if they were willing to assume that the probability of a major disruption was less than 1 percent per year. Perhaps Schlesinger would have taken this approach if his analysts had not focused his attention on the apparent discrepancy between DOE and OMB and he had felt more confident of the program's implementation record.

In early January OMB began pressuring DOE to participate in another joint study of the economic benefits and costs of implementing the billion barrel goal. The director of the Office of Contingency Planning was reluctant to begin yet another study of the size issue, especially as a joint effort with OMB. The office was heavily involved in analyses related to the Iranian crisis and was attempting to complete work on the SPR distribution plan. In an internal DOE memorandum, the director argued against another joint study:

> **Problem With OMB Re SPR Size**
>
> *DOE will establish its work plan* after considering OMB proposals. DOE must be responsible for use of its resources.
>
> *No joint effort* with OMB. We will do our work plan and inform OMB of results and DOE position. Experience last year with a "joint effort" was unacceptable. OMB staff continuously attempting to manage the studies, but not responsible for the results or costs. Then OMB went to the President without any previous consultation with DOE.
>
> *Do not expect to reach agreement on all assumptions of "facts."* In the past we have had irreconcilable differences with OMB staff. We see no reason for that to change. Most of the differences about assumptions cannot be resolved by analysis because they are projections or predictions of potential future considerations. No reason to expect OMB staff to agree to any assumption that might support an SPR size larger than 750 MMB.[27]

Despite this protest by the Office of Contingency Planning, DOE agreed to a joint study with OMB to be completed in September for consideration in fiscal year 1981 budget proposals.

Personnel from other offices within Policy and Evaluation, the Energy Information Administration, and the Office of International Affairs worked with members of the Office of Contingency Planning on the study. In all, about twenty person-months of professional staff time from DOE were expended over a four-month period beginning in June. Consultants were paid approximately $80 thousand, primarily for simulations using the Wharton and DRI macroeconomic models.

Over the summer representatives from DOE, OMB, and CEA met every week or two to discuss results from the macroeconomic simulations of disruptions. Although all parties initially agreed to a basic set of assumptions for simulations, the OMB and CEA representatives continually suggested changes. These changes were very frustrating for the DOE representatives who had to implement them, particularly since it appeared that the OMB representatives were looking for ways to make disruptions appear less costly and, hence, to make the SPR appear less valuable.

DOE completed a first draft of the study report in October.[28] The analysis divided the SPR into two components: a primary reserve that would be used to assure the "essentials of survival" during a severe disruption and a secondary reserve that would be used to reduce the economic losses caused by disruptions.

The analysis drew on other studies to estimate that U.S. petroleum use could be cut by 25 percent during an emergency without jeopardizing health, safety, or peacetime national security. It was also assumed that industry could be induced to draw down its reserves by 125 million barrels during a severe disruption. Under these assumptions, the report argued that a primary reserve of at least 550 million barrels was needed to preserve services such as food production and distribution, home heating, medical care, police and fire protection, and national defense during severe disruptions. Examples of severe disruptions considered in the report were a 50 percent reduction in exports from OAPEC and Iran for nine months and a 75 percent reduction in Persian Gulf exports for six months with a 50 percent reduction for the following six months, and a Persian Gulf closing in conjunction with simultaneous prosecution of a major conventional war by the United States. The report acknowledged that the probabilities of these events were low. It argued, however, that they were sufficiently likely and their consequences so severe as to justify a minimum-size primary reserve of 550 million barrels.

The desirable size of the secondary reserve was determined by finding the size for which marginal costs equaled marginal expected benefits under the assumptions being considered. Drawdown benefits were measured from functions relating U.S. shortfalls to GNP losses that were derived from the Wharton and DRI macroeconomic model simulations under various assumptions about import levels and the existence of crude oil and petroleum product price controls in the United States. Eight disruption scenarios were analyzed. Each scenario was assigned a probability of occurrence in the eight-year period between 1982 and 1990. The probabilities ranged from 80 percent for a "small single country interruption" to 1 percent for the severe Persian Gulf scenario. It was assumed that the primary reserve would be used in disruptions involving losses of greater than one billion barrels per year. An industry drawdown of 125 million barrels was assumed.

The report concluded that a secondary reserve of at least 450 million barrels was desirable on economic grounds. It asserted that even relatively small disruptions, such as a single medium-size producer being shut down for six months, would make it economically attractive to have a secondary reserve of 450 million barrels. It suggested that the most economically desirable secondary reserve size ranged from 1.6 billion to 6 billion barrels. In other words, it argued that a total reserve of larger than one billion barrels was justified.

OMB was far from pleased with the draft report. OMB was preparing to argue not only against Phase IV, but also against Phase III on the grounds that new capacity should not be added while existing capacity was not being filled.[29] In light of the long lead time needed for storage facilities, the OMB position implied that the then current tight market conditions caused by the reduction in Iranian production would prevail for five or six years. Nevertheless, the low credibility of the SPR Program in the wake of revelations about cost overruns and schedule delays during the previous year encouraged OMB to try to cut Phase III.

OMB raised four major objections to the draft report. First, the use of the assumed probabilities of disruptions was attacked as arbitrary. Although DOE believed the assumptions to be conservative, they had no way to defend the particular probabilities that were used. Previous attempts to get the National Security Council to provide estimates of probabilities had failed. OMB preferred calculation of break-even probabilities instead.

Second, OMB wanted the GNP-loss function to be estimated using more accommodating fiscal and monetary assumptions than had been initially agreed upon. These policies would make the real GNP losses caused by oil shortfalls appear less severe by reducing declines in aggregate demand. At the same time, however, inflation rates, which were not part of the economic loss being measured, would increase. Because both the DRI and Wharton models are demand-driven, they are really not satisfactory tools for measuring supply-induced price shocks. Modelers must make adjustments, which are often arbitrary, to reflect supply-side effects.[30] Such adjustments become more difficult when extreme monetary and fiscal policies are assumed; the corresponding results are more suspect.

Third, OMB objected to the assignment of a salvage value to SPR oil. DOE assumed that if no disruption occurred during the study period, the oil accumulated in SPR would be sold at the world price. Because the real price of oil was expected to rise over the study period, the sale of oil at the end of the period would greatly reduce program costs. DOE also offset program costs from the revenue obtained from the sale of SPR oil during disruptions. OMB opposed the inclusion of either of these revenues in the calculation of costs and benefits.

Fourth, OMB argued that SPR drawdowns would not replace lost imports barrel for barrel. DOE had assumed that simultaneous drawdowns of stocks by other members of the International Energy Agency (IEA) would prevent SPR drawdowns from displacing U.S. imports. If other IEA members did not draw down their reserves in coordination with the United States, a fraction of the SPR oil would in effect "leak out" of the United States in the form of displaced imports. The leakage problem was the most valid of the four objections raised by OMB.

DOE realized that Phase III was threatened and decided to revise the analysis to find the break-even probability that would justify a secondary reserve of 200 million barrels (or a total reserve of 750 million barrels) under a set of assumptions that OMB would find difficult to attack. The analysis focused on a moderate-size disruption causing a 625 million barrel shortfall to the United States over one year. It was assumed that 125 million barrels of the shortfall would be offset by drawdowns of private sector stocks and 300 million barrels by drawdown of the primary reserve. The benefits of the secondary reserve were measured in terms of GNP losses avoided by offsetting a fraction of the remaining shortfall and compared to the costs of developing the secondary reserve.

A new GNP-loss function was derived, assuming accommodating monetary and fiscal policies were employed. At the suggestion of CEA, it was based on the 1985 Wharton macroeconomic forecasts with the high vulnerability price path (low price path with correspondingly higher import levels). CEA recommended use of the high vulnerability price path because the low vulnerability path suggested structural changes in the economy inconsistent with the existing equations in the model. Problems were encountered in using the low vulnerability price path in 1990 scenarios: the Wharton and DRI models gave widely inconsistent results. The GNP-loss function represented the discounted value of three-year reductions in real GNP. The GNP losses were counted as costs along with facility and oil purchase costs.

A short-run oil price elasticity of demand of −.20 was assumed for the U.S. and −.25 for the rest of the world. In a fully unconstrained market, these elasticities would result in a leakage to other consuming countries of three of every four barrels withdrawn from the SPR. Because the IEA sharing agreements were assumed to be in effect, a leakage of only one barrel for every four was assumed. Less elastic demand would have led to a smaller estimate of leakage and a smaller break-even probability. A number of other assumptions leading to a higher break-even probability were made: no price ratchet following an interruption, no credit for revenue from sales of SPR oil during disruptions, and an across-the-board reduction of 25 percent of benefits due to inefficiencies in SPR use.

Under these assumptions, the break-even probability was calculated to be 25 percent for the eight-year study period or about 3.5 per year.[31] The report went on to argue that because a disruption of the magnitude being considered was much more likely than the break-even probability, a reserve of at least 750 million barrels was justified. By the time the report was completed, however, DOE had decided not to appeal OMB's rejection of its request for Phase III funding to the president.[32] Additionally, the DOE under secretary agreed to yet another joint OMB/DOE size study to be completed during preparation of the administration's fiscal year 1982 budget.

Resolution of the Phase III Controversy

While the Contingency Planning Office (renamed the Emergency Response Planning Office in January 1979) was struggling with OMB over the assumptions to be used in the joint study, two other offices also under the assistant secretary for Policy and Evaluation conducted studies dealing with the size issue. The Office of Natural Gas and Integrated Analysis in DOE began a study that led to estimates of optimal stockpile sizes in combination with tariffs and other import reduction policies.[33] Although the general approach was to look at both macroeconomic costs and benefits, the benefits of stockpile drawdowns during disruptions were measured in terms of avoided GNP losses only. The costs of increased oil prices caused by acquisition of oil were included. It was assumed that never more than half of the stockpile would be drawn down. For a set of representative disruptions and their postulated probabilities of occurrence, the minimum optimal stockpile size was calculated to be six billion barrels. Perhaps because stockpiling was only one of several topics addressed in this long and diffuse study, it did not play an important role in the policy debate.

A second study, which did turn out to be important in the debate with OMB, was initiated for the Office of Oil by Thomas J. Teisberg of the Massachusetts Institute of Technology. He realized that optimal strategies for acquisition and drawdown decisions for the SPR could not be determined without a more systematic treatment of uncertainty about the timing, severity, and duration of disruptions. He reasoned that if assumptions were made about the probabilities of occurrence of various states of the market, then the problem could be handled in the decision analysis framework. To solve the decision analysis problem, he developed a computerized dynamic programming model. This model produces a contingency table for each year of the study period indicating the drawdown and acquisition decisions that maximize the present value of expected net benefits (measured in terms of social surplus changes resulting from changes in the world oil price) for each possible entering stockpile size and state of the market. The stockpile size eventually reaches a point where it is no longer optimal to make additional purchases. This plateau level is an indication of the "optimal" stockpile size for planning purposes.

In December 1979 the Office of Oil staff completed a major study of acquisition and drawdown strategies using the Teisberg model.[34] Four states of the market were included: slack market in which SPR purchases of up to 5 percent of OPEC production could be made without causing a rise in the world price; tight markets in which each 100 thousand barrels per day of SPR purchases increase the world oil price by $1.50 per barrel; minor disruptions corresponding to 10 percent of OPEC production for one year; and major disruptions corresponding to cutbacks of 25 percent of OPEC for one year. For each run of the model, assumptions were made about the probabilities of moving from each market state to each of the other market states, the path of U.S. import levels, and the short-run elasticity of demand for imported oil.

Assuming slack markets to be available 50 percent of the time, the smallest plateau level estimated was 800 million barrels. It resulted under assumptions corresponding to a minor disruption occurring on average once every 12 years, and a major disruption occurring on average once every 20 years, a U.S. import level of 4.3 million barrels per day in 1990, and a short run elasticity of demand for imported oil of −.52.[35] More pessimistic assumptions about the occurrence of disruptions, import levels, and import demand elasticities yielded plateau sizes as large as 4.4 billion barrels. The report implied that a one billion barrel SPR was justified on economic grounds and recommended a fill rate of 550 thousand barrels per day.

In December 1979 the DOE under secretary reached tentative agreement with OMB and CEA to set aside up to one million dollars of the FY 1980 budget for a study to support future decisions on the ultimate size, development schedule, and use of the SPR. It was proposed that DOE, OMB, and CEA jointly agree on the methodology and assumptions to be employed by a "qualified independent contractor."

In a memorandum to the under secretary, the assistant secretary for Policy and Evaluation recommended that the study be done in-house through extension of the analysis on acquisition and drawdown strategies completed by the Office of Oil. The assistant secretary warned that agreement on assumptions would be difficult and suggested that the analysis be performed under a range of assumptions. He suggested that the Office of Oil methodology could be used to handle all the issues raised except the effects of various monetary and fiscal policies to reduce the economic losses during disruptions, and he implied that investigation of these macroeconomic issues were not an appropriate component of the SPR study.

Personnel from the Office of Oil, OMB, and CEA met in February 1980 to discuss a critique of the Teisberg model prepared by the CEA staff. The assumptions made about the underlying paths of the world oil prices, the short-run and long-run price elasticities of U.S. import demand, and the price elasticity of demand in the world market were questioned as arbitrary. The DOE analysts responded that sensitivity of results to variations in these assumptions could be easily determined through additional runs of the model. Objections were raised to the use of social surplus as a benefit measure because it does not capture the macroeconomic losses caused by severe disruptions. The DOE analysts agreed that the social surplus approach probably underestimates the costs of disruptions and hence the benefits of the SPR but indicated the Teisberg model could be modified to measure benefits in terms of any GNP or other loss function that depended upon changes in the price of oil. Finally, an objection was raised to the use of expected values on the grounds that it implied a linear loss function and risk neutrality. The DOE analysts agreed that it implied risk neutrality but argued that expected value calculations did not imply a linear loss function. The social surplus loss function was not linear in price, and any alternative loss function could be used in its place.

The DOE analysts believed they had successfully argued that the Teisberg model could be modified to satisfy the objections that had been raised. But at the end of May another set of OMB and CEA criticisms arrived with an indication that a proposal for an alternative methodology would follow. The memorandum reiterated the previously raised objections and a number of new ones.

At the conceptual level, CEA worried that the use of expected values could "seduce the decision maker" into considering only net benefits to the exclusion of the "possibly calamitous effect on the economy and national security should a major disruption occur in the absence of adequate SPR insurance." The use of a transition matrix of probabilities of going from one state of the market to another was labeled as "so complex and theoretical as to have no meaning to policymakers." Objections were again raised to the use of social surplus as a benefit measure because it failed to take account of macroeconomic effects. The model was criticized for not taking account of limitations in the rate at which storage capacity can be added and filled. It was also criticized for not taking account of the timing of expenditures for facility development.

The OMB objections to particular assumptions appeared to be intended to show that the Office of Oil analysis overestimated the value of SPR. Most of these, such as the insistence that a real discount rate of 10 percent rather than 8 percent be used, would not significantly alter the Office of Oil results. Those of most importance were related to the size and likelihood of disruptions considered. OMB argued that the shortfall sizes assigned to minor and major disruptions were overly pessimistic because they were beyond the range of historical experience. It was also suggested that the assumed probabilities implied excessive pessimism about the frequency of disruptions.

The Office of Oil responded to the conceptual objections with a general defense of the expected value approach. It was indicated that the storage capacity constraints could be incorporated and again noted that a GNP or other loss function could be substituted for consumer surplus. As to the assumptions about size and frequency of disruptions, the Office of Oil reported that almost all other reviewers of the analysis, both inside and outside of DOE, felt the assumptions were too optimistic. In fact, several reviewers recommended that a market state be included to represent a closing of the Persian Gulf, a 65 percent loss of OPEC supply.

OMB and CEA rejected the Teisberg approach. In early July the assistant secretary for Policy and Evaluation received a memorandum outlining the joint study that OMB and CEA proposed be completed by mid-September.[36] The study constituted yet another effort to measure the economic costs of disruptions through use of the DRI and Wharton macroeconomic models. Three disruptions corresponding to losses of Iranian and Iraqi production for one year, Saudi Arabian production for one year, and 75 percent of Persian Gulf production for one year, would be modeled for SPR sizes of zero, 538 million, 750 million, and one billion barrels. In all, thirty-two basic scenarios would be analyzed with each of the macromodels. For the twelve quarters following the onset of each disruption, a large number of macroeconomic variables, ranging from real GNP to automobile production, would be followed. Comparisons of disruption scenarios with the various SPR sizes would be made in terms of the price of oil, the overall price level in the economy, the level of output, the unemployment rate, and the federal budget. These results would be made available to some "responsible policymaker" (the president?) who would balance the various costs and benefits to arrive at a decision about the appropriate size of the SPR.

The Office of Oil staff looked upon the proposal with horror. Not only had the basic approach been employed before by DOE, both with and without OMB participation, but a similar analysis had just been completed by the Congressional Budget Office.[37] Hence the anticipated frustration of trying to reach a consensus about assumptions would purchase little in the way of new insights or firm conclusions. The thought of presenting detailed descriptions of the behavior of macroeconomic variables to the president or even the secretary seemed ridiculous, especially if they were then expected to subjectively take account of all the aspects of uncertainty. Further, the staff held the view that it was wishful thinking to believe that the demand-driven macroeconomic models would provide reasonable predictions of the effects of oil supply shocks occurring eight to ten years in the future. In fact, the staff was surprised when CEA member George Eads endorsed the study because during the 1979 study he had raised questions about the validity of using the econometrically derived macromodels for predictions of the effects of disruptions occurring far in the future.

A compromise of sorts was reached at a July 10 meeting of the assistant secretary for Policy and Evaluation (DOE), the assistant director for Natural Resources, Energy, and Science (OMB), a council member of CEA, and their respective staffs. The assistant secretary began by protesting the late date of the OMB/CEA proposals. He next reviewed the merits of the analysis already completed by his staff and expressed a willingness to accommodate OMB and CEA suggestions in its extension. In particular, he requested that CEA provide what they believed to be an appropriate loss function for use in the Teisberg model. When pressed on the commitment made by the former under secretary for a joint study, the assistant secretary said his office would provide contractor support of about $100,000 to OMB and CEA for the study on the understanding that DOE would present its own analysis as part of the budget request for FY 1982. When asked about staff support, he said he would not commit the sixteen man-months that he believed, on the basis of the effort the previous year, would be needed. He did agree to provide the two man-weeks of staff support OMB said it needed on the condition that OMB and CEA would take responsibility for all the assumptions used in the analysis. This agreement was followed with one exception; the next day OMB extracted a commitment of four man-months of DOE staff time because its initial estimate had been too low.

Office of Oil personnel provided energy related projections as requested by OMB/CEA and kept track of the assumptions being used in the macromodel runs. Within a short time, it became apparent that the assumptions made by CEA and passively agreed to by

OMB were going to yield results even more supportive of larger SPR sizes than the 1979 study. OMB and CEA never completed the proposed study. They did not even provide DOE with a report of the results that were derived. When the Office of Oil requested a copy of the final report, the deputy associate director of the Special Studies Division for Natural Resources, Energy and Science, who directed the study for OMB, indicated that it was part of OMB's internal budget review and therefore would not be made available for distribution.

In the meantime, the Office of Oil had modified the Teisberg model to focus explicitly on the Phase III question. The results were presented in October at the OMB hearing on the SPR budget. Perhaps because the macromodel analysis had not turned out as they had expected, personnel from OMB's Special Studies Division for Natural Resources, Energy, and Science finally expressed an interest in having analysis from the Teisberg model. The analyses were completed with the understanding that DOE did not necessarily endorse the assumptions that OMB requested.

By mid-November it became clear that OMB would not oppose funding in the administration's FY 1982 budget request for implementation of Phase III. Did analysis finally carry the day? Perhaps. In light of the extended fight over Phase III, however, more plausible explanations can be offered. Most likely, the OMB staff anticipated that the assistant secretary for Policy and Evaluation, who enjoyed the confidence of the secretary, would recommend that the issue be taken to the president if Phase III funding was not included in FY 1982 budget proposal. Unlike the previous two years, it appeared that the DOE leaders were confident of their analysis and willing to take OMB to the mat. At the same time,the SPR program had begun to reestablish its managerial credibility, undercutting implied arguments that it could not handle implementation of Phase III. Finally, perhaps with excessive hindsight, one cannot help but wonder if the presidential election was not a factor. As it turned out, President-elect Reagan appointed David Stockman, a strong congressional supporter of the SPR, as director of OMB. It may be that the career staff at OMB anticipated this possibility and did not wish to be seen by the new administration as being responsible for further delay of the SPR program. In fact, during the early months of the Reagan administration the same OMB staffers, who had led the fight against expansion of the SPR throughout the Carter administration, directed a joint study with DOE of options for accelerating Phase III. However, by the following year they were again seeking delay of Phase III.

The Size Issue in Perspective

Is the rapid development of a billion barrel SPR desirable in terms of economic costs and benefits? In 1977, when the goal was established, it was based on national security rather than analytically supported economic considerations. Beginning with the studies in 1978, however, a preponderance of analytical results have supported the desirability of starting development of a billion barrel reserve. These results reflect a variety of methodological approaches and are robust over the fairly wide range of assumptions. In addition to the studies already discussed, analyses by Henry Rowen and John Weyant of Stanford University,[38] Hung-Po Chao and Alan Manne,[39] and James L. Plummer of the Electric Power Research Institute[40] seem to justify on economic grounds immediate development of a reserve of one billion barrels or larger.

Although most of the early analyses dealt with the question of the appropriate ultimate size of the SPR, the question of greater policy relevance is how much capacity should we start to build today. The Teisberg model has been modified to answer this question through the introduction of constraints on the availability of storage capacity.[41] This analysis is summarized in table 5.

Under optimistic assumptions about future conditions in the world oil market, accelerated completion of the billion barrel reserve by 1986 with storage facility costs of $6 per barrel or less offers greater net economic benefits than slower development schedules assuming leached salt dome cavern storage at $4 per barrel. Under more pessimistic assumptions about the future, completion of a billion barrel reserve by 1986 would be desirable even if storage costs were as high as $12 per barrel, close to the estimated costs of constructing above-ground steel tank storage. In other words, more costly but faster construction of steel tank storage would be preferable to the less costly but slower construction of salt dome capacity.

Several noneconomic benefits further strengthen the case of a billion barrel reserve. A larger SPR gives the United States greater flexibility in dealing with political and military conflicts in the Per-

Table 5
Net Economic Benefits of the SPR[1]
(billions of 1980 dollars)

	OPTIMISTIC EXPECTATIONS[2]	*PESSIMISTIC EXPECTATIONS*[2]
Phases I-II (550 mmb by 1986)	17.1	68.2
Phases I-III (750 mmb by 1989)	17.6	73.1
Phases I-IV (1,000 mmb by 1991)	17.5	76.2
Phases I-IV Accelerated (1,000 mmb by 1986)		
Storage at $6/barrel	18.2	83.0
Storage at $12/barrel	14.5	79.3

SOURCE: Glen Sweetnam, "Reducing the Costs of Oil Interruptions: The Role of the Strategic Petroleum Reserve," Presented at the ORSA/TIMS Joint National Meeting at Colorado Springs, Colorado, November 10, 1980.

[1] Present discounted value (10 percent real discount rate) of all microeconomic costs and benefits of the SPR based on Office of Policy and Evaluation SPR Model.

Assumptions: U.S. and world short run price elasticity of demand for imported oil of -0.25; world base price rises at a real rate of 3 percent per year to a backstop price of $65 per barrel (1980 dollars); and U.S. imports of 7 million barrels per day until 1990, declining to 5 million barrels per day by the year 2000.

[2] Expectations about time spent in each of five world market states:

1. slack market—oil purchases possible without increase in world price
2. tight market—oil purchases increase world price.
3. minor disruption—2 million barrels per day worldwide for one year
4. moderate disruption—6 milion barrels per day worldwide for one year
5. major disruption—12 million barrels per day worldwide for one year

Average Time Spent in Each State (in percent)

	Slack	Tight	Minor	Moderate	Major
Optimistic	50	25	20	5	0
Pessimistic	31	33	20	10	5

sian Gulf and other oil producing regions. It may deter embargoes or other politically motivated disruptions of the world oil market. Its use during disruptions may help diffuse political pressure for economically inefficient energy price controls and allocation programs, which would be likely to linger long after the disruption has ended.

Regret analysis, which considers worse case consequences of decisions, also suggests a large reserve is desirable. What would be the costs to the United States if a billion barrel reserve is developed and U.S. vulnerability ends without a disruption having occurred? Assume that 100 million barrels are added to the reserve in each of the next ten years at a real cost of $40 per barrel for oil and $4 per barrel for storage facilities and the stockpile is sold at $40 per barrel in the eleventh year. The direct budgetary outlays would be over $44 billion. The present value of real expenditures for developing the reserve would be $29.8 billion, assuming a real discount rate of 10 percent. The present value in real dollars from sale of the oil at program termination, however, would be $15.4 billion, yielding a net present value of real costs of $14.4 billion. This amount, which would be smaller if the real price of oil increased over the decade, represents the premium paid for insurance that never pays a direct benefit. On the other hand, what would be the consequences of not having a billion barrel SPR in the face of a one year loss of Saudi Arabian production or closure of the Persian Gulf? With no SPR the real GNP losses would probably be between $200 billion and $300 billion, inflation would increase by 15 to 25 percentage points, and the unemployment rate would increase by about 2 percentage points. Even with a reserve of 750 million barrels, the regret associated with not having an additional 250 million barrels is probably a minimum of $20 billion dollars.[42]

It thus appears from several perspectives that a billion barrel SPR is desirable. The current schedule for the development of storage capacity, however, greatly increases the uncertainty of this conclusion. As now planned, Phase III storage will not be completed until 1989. If phrase IV were to be made available through the existing SPR program, it is unlikely that it would be completed until well into the 1990s. The large oil price increases of recent years have already resulted in reductions in oil consumption through changes in capital stock. The price increases, coupled with the eliminations of price controls in the U.S., have also encouraged private firms to hold somewhat larger inventories. If these trends were to continue, the value of Phrase IV would be lower than current studies suggest. Of course, a stabilization or reduction of real oil prices could reverse these trends. Because the benefits are larger and more certain and the sooner oil is actually in storage, higher expenditures for the acceleration of Phases III and IV are likely to be justified in terms of expected net benefits. Because it would not necessarily have to be filled, the regret associated with beginning Phrase IV construction now and later deciding it to be unnecessary is on the order of one billion dollars—a large but not unreasonable insurance premium.

If Phases III and IV are justified, how should we view OMB efforts to delay their implementation? OMB demands for better analysis of the economic costs of the SPR expansion in 1977 should probably be viewed favorably. Even though the president had already made a commitment to the billion barrel goal, it is important that dissenting voices be raised when decisions appear to have been based on incomplete analysis. Once the 1978 study was completed, however, it should have been apparent to OMB personnel that the presidential decision, made primarily in national security grounds, was also justified in terms of economic costs and benefits. Further attempts to delay implementation through what one participant called the "paralysis of analysis" seems misguided from the "public interest" point of view.

The obvious cost of the OMB tactics was the two-year delay in the start of Phase III. The demands for repetition of studies might be viewed in a more positive light if OMB had made greater methodological contributions; instead, much of their effort represented attempts to find combinations of assumptions that would support their position. More generally, one can wonder if SPR implementation problems might have been more easily overcome with greater OMB support in such areas as obtaining environmental permits, developing management control systems, and designing alternative financing methods.

In conclusion we consider two questions related to the nature of the relationship between DOE and OMB: Why did the DOE hierarchy back down on the size issue during the FY 1980 and FY 1981 budget cycles? How can we explain the willingness of the DOE hierarchy to hold its ground on the FY 1982 budget?

DOE must deal with OMB on numerous budget issues. To maintain its credibility, it must be selective in choosing the issues it takes to the president. Although Secretary Schlesinger did take the size issue to the president in 1977, other issues gained in relative importance within the newly created DOE. In a study of the first two years of DOE contingency planning, which includes the SPR as its most important element, the Committee on Governmental Affairs of the U.S. Senate recently concluded:

> In short, contingency planning suffered after the formation of DOE for three major reasons: preoccupation with other matters of substance, organization, and personal "turf" interest; lack of leadership by top officials inside and outside the Department; and, ironically, the reality that no major interruption of oil supply had occurred.[43]

The oil price rises in 1979 following the Iranian revolution did bring home the reality of major oil supply disruptions. By that time, however, Schlesinger's influence in the administration had greatly weakened, culminating in his resignation as secretary in July 1979.

With respect to the SPR specifically, the problems encountered in implementing Phases I and II were embarrassing to the top DOE officials. Secretary Schlesinger himself was forced to publicly admit that the SPR had suffered from "bad management."[44] Vulnerability to such charges, if not real doubts about the capabilities of the SPR Office, probably were important factors in the hesitancy of the DOE leadership to go to the mat with OMB over Phases III and IV. By early 1980, however, much progress had already been made in improving the management capabilities and restoring the credibility of the SPR Office. Nevertheless, the new DOE secretary did not press for Phase III funding in 1979.

The improved credibility of the SPR Office and the harsh reminder of our vulnerability to market disruptions provided by the Iranian cutbacks of production were general factors in the greater willingness of DOE leadership to press the Phase III fight in 1980. An explanation can also be found in the actions of two individuals who played the role of "policy entrepreneurs."[45] One was the director of the Office of Oil, who early in his tenure spotted the SPR issue as important, He encouraged Teisberg to work on the problems of optimal acquisition and drawdown. When Teisberg developed a methodology that effectively dealt with the questions related to uncertainty, the director was able to convince the assistant secretary for Policy and Evaluation that the analysis was sound and its results important. The assistant secretary then played the role of policy entrepreneur in confronting OMB with the analysis and winning the support of the secretary.

The assistant secretary had several resources that made him an effective player in the policy process. He enjoyed a reputation for giving advice largely on the basis of analytical rather than political considerations.[46] He was probably the most influential advisor to the secretary. He invested effort in understanding analysis and therefore had confidence in its results. These factors put him in a strong position for dealing with OMB and even allowed him to limit somewhat the amount of "rolling in the mud" his staff had to do with the OMB staff.[47]

Notes

1. Executive Office of the President, "The National Energy Plan," April 29, 1977 (Washington, D.C.: Government Printing Office, 1977), p. 60.

2. Strategic Petroleum Reserve Office, "Strategic Petroleum Reserve Plan Amendment No. 2 (Energy Action No. 1): Expansion of the Strategic Petroleum Reserve," March 1978, DOE/RA-0032/2.

3. U.S. Department of Energy, "National Energy Plan II," May 1979, p. 38.

4. For a history of OMB, see Larry Berman, *The Office of Management and Budget and the Presidency 1921-1979.* (Princeton, N.J.: Princeton University Press, 1979); and Hugh Heclo, "OMB and the Presidency—The Problem of Neutral Competence," *The Public Interest*, no. 38 (Winter 1975), pp. 80-98.

5. For a comprehensive bibliography, see The Aerospace Corporation, "SPR Size Studies Review," prepared for the Strategic Petroleum Office, U.S. Department of Energy, November 15, 1980.

6. National Petroleum Council, "Emergency Preparedness for Interruptions of Petroleum Imports into the United States," September 1974.

7. Federal Energy Administration, "Project Independence Report," November 1974, pp. 385-86.

8. Interview with Robert L Davies, October 8, 1980.

9. The Institute for Defense Analyses study measured benefits in terms of consumer surplus. It recommended the creation of a reserve of 4.38 billion

barrels. For a summary, see Robert E. Kuenne, Gerald F. Higgins, Robert J. Michaels, and Mary Summerfield, "A Policy to Protect the U.S. Against Oil Embargoes," *Policy Analysis* 1, no. 4 (Fall 1975): 571-97.

10. For example, a major review of methodological issues was provided by J. Philip Childress, Glenn Coplan, and Lowell Goodhue, "Strategic Storage Program Preliminary Cost/Benefit Analysis," Draft, FEA Office of the Assistant Administrator for Policy and Analysis, August 29, 1975.

11. J. Philip Childress, "Description: Strategic Petroleum Reserve Cost/Benefit Model," FEA, Office of the Assistant Administrator for Policy and Analysis, August 20, 1975.

12. J. Philip Childress, "Optimal Stockpile Size Based on Total 15-year Life Cost Assuming a Single Import Disruption," Discussion Paper, FEA, Office of the Assistant Administrator for Policy and Analysis, July 8, 1975.

13. Randall G. Holcombe, "A Method for Estimating the GNP Loss from a Future Oil Emargo," *Policy Sciences* 8, no. 1 (June 1977):271-34.

14. Strategic Petroleum Reserve Office, "Strategic Petroleum Reserve Plan," December 15, 1976, Chapter 2 and Appendix I.

15. This discussion is based primarily on interviews with Carlyle Hystad, then head of analytical services for the SPR Office. Interview, September 15 and 16, 1980.

16. In March 1977 the SPR Office completed an analysis suggesting that a billion barrel reserve would be cost-effective against a severe disruption of the type that might be encountered during wartime. Because the SPR Office analysts believed such a scenario unlikely, they recommended a 750 million barrel reserve as cost-effective against the range of likely disruptions. Carlyle E. Hystad, "Estimating Appropriate Reserve Size: 750 and 1000 MMB," Federal Energy Administration, Strategic Petroleum Reserve Office, March 24, 1977. At about the same time Egan Balas of Carnegie-Mellon University presented his analysis of the deterrent benefits of the SPR. He concluded that embargoes become unattractive for producers when the U.S. has reserves of between 500 million to 1,300 million barrels, depending upon levels of U.S. imports from OAPEC. Egan Balas, "Recommendation for the Overall Size of the SPR," Appendix V of Review of the Strategic Petroleum Reserve Plan, U.S. Senate Hearings before the Committee on Interior and Insular Affairs, February 4, 1977. pp. 491-508.

17. James L. Cochrane, "Carter Energy Policy and the Ninety-fifth Congress," in Craufurd G. Goodwin, ed., *Energy Policy in Perspective* (Washington, D.C.: The Brookings Institution, 1981), p. 555. Cochrane describes the other agencies as "waiting for the Delphic utterance from the Old Executive Office Building." p. 356.

18. This paragraph is based on an account given to me by Carlyle Hystad, who attended the meeting. Interview. September 16, 1980.

19. Strategic Petroleum Reserve Office, "Strategic Petroleum Reserve Plan Amendment No. 2 (Energy Action DOE No. 1): Expansion of the Strategic Petroleum Reserve," March 1978.

20. U.S. Congress, House, "Cost/Benefit of the Strategic Petroleum Reserve," Hearings before the Subcommittee on Energy and Power of the Committee on Interstate and Foreign Commerce, 95th Cong., 2d sess., February 16, 17, and 23, 1978, pp. 152–53.

21. Memorandum on "Oil Supply Disruption Contigency Plan Task Force," from Eliot Cutler, Associate Director for Natural Resources, Energy, and Science, OMB, to Al Alm, Assistant Secretary for Policy and Evaluation, DOE, February 23, 1978.

22. David Couts et al., "Economic Analysis of Petroleum Supply Interruption Contingency Actions, Appendices to Volume I: Simulation Specifications and Results," Science Applications, Incorporated, September 15, 1978; and William P. Curtis, "Macroeconomic Effects of Petroleum Supply Interruptions: Volume I," Department of Energy, Energy Information Administration, Office of Intergrated Analysis, March 1979. (Results from the EIA study were available in the fall of 1978).

23. In a July 3, 1978, letter to Secrerary Schlesinger, OMB Director James T. McIntyre, Jr., expressed concern that inadequate progress was being made in completing analyses necessary for consideration of funding for the billion barrel reserve in FY 1980.

24. Memorandum on "Follow-up questions—DOE justification for the fourth 250 mmb of storage by 1985," from Ken Glozer, Special Studies Division for Natural Resources, Energy, and Science, OMB to George McIsaac, assistant secretary for Resource Applications and Al Alm, assistant secretary for Policy and Evaluation, October 20, 1978.

25. Letter to Eliot Cutler, associate director for Natural Resources, Energy and Science from George S. McIsaac, assistant secretary for Resource Applications, and Alvin L. Alm, assistant secretary for Policy and Evaluation, November 1, 1978.

26. U.S. Department of Energy, Strategic Petroleum Task Force, "DOE Analysis of the Need for the Fourth 250 MMB of the Strategic Petroleum Reserve and DOE Comments on OMB's Position," December 18, 1978.

27. Internal DOE Memorandum, Carlyle Hystad, January 22, 1979.

28. Strategic Petroleum Reserve Task Force, "DOE Analysis of the Appropriate Size of the Strategic Petroleum Reserve," draft, October 11, 1979.

29. Statements issued from the Tokyo Conference in June suggested that the U.S. would not resume filling the SPR until the market stabilized. Tokyo Communique, "Joint Declaration of Tokyo Summit Conference," June 29, 1979. In *Public Papers of the Presidents of the United States, Jimmy Carter,*

Book II, June 23 to December 31, 1979 (Washington, D.C.: U.S. Government Printing Office, 1980).

30. For a discussion of the macroeconomic models in layman's terms, see "Where the Big Econometric Models Go Wrong," *Business Week*, March 30, 1981, pp. 71–72, 77.

31. Glenn Coplan et al., "DOE Analysis of the Appropriate Size of the Strategic Petroleum Reserve," Working Draft, U.S. Department of Energy, Assistant Secretary for Policy and Evaluation, Office of Emergency Preparedness, November 30, 1979.

32. On August 20, 1979, DOE announced cancellation of the turnkey solicitation which would have provided Phase III storage through fixed price contracts with private firms. DOE requested funding for FY 1980 and FY 1981 to permit expansion of government developed storage. OMB rejected these requests.

33. Jerry Blankenship, Mike Barron, Joseph Eschbach, Linsay Bower, and William Lane, "The Energy Problem: Costs and Policy Options," Staff Working Paper, U.S. Department of Energy, Assistant Secretary for Policy and Evaluation, Office of Oil, May 23, 1980.

34. Glen Sweetnam et al., "An Analysis of Acquisition and Drawdown Strategies for the Strategic Petroleum Reserve," Draft, U.S. Department of Energy, Assistant Secretary for Policy and Evaluation, Office of Oil, December 17, 1979.

35. The elasticity of demand for imports (e_I) is related to the elasticity of demand (e_D) and the elasticity of supply (e_S) by the following formula:

$$e_I = (D(P)/I(P))e_D - (S(P)/I(P))e_S$$

where D(P) is the quantity demanded, I(P) the import level, and S(P) the quantity domestically supplied at price P. For example, if D(P) = 18 mmb/d, D(S) = 10 mmb/d, I(P) = 8 mmb/d and the supply elasticity equals 0.056, the −0.52 import elasticity corresponds to a demand elasticity of −0.20.

36. Memorandum on "SPR Benefit/Cost Study," from Kitty Schirmer and George Eads to Bill Lewis, July 1, 1980.

37. Barry J. Holt and Mark Berkman, "An Evaluation of the Strategic Petroleum Reserve," U.S. Congress, Congressional Budget Office, Natural Resources and Commerce Division, June 1980.

38. Rowen and Weyant use a model of energy-economy interactions developed by James L. Sweeny of Stanford University. They find that a billion barrel SPR is justified by a 2.7 percent probability of a Saudi Arabian shutdown for one year. Henry Rowen and John Weyant, "The Optimal Strategic Petroleum Reserve Size for the U.S.?" Stanford University, International Energy Program Discussion Paper, October 1979.

39. Hung-Po Chao and Alan S. Manne, "Oil Stockpiles and Import Reductions: A Dynamic Programming Approach," Electric Power Research Institute, Palo Alto, California, October 1980.

40. Using the model developed by Hung-Po Chao and Alan Manne, Plummer reports a stockpile plateau size of about two billion barrels under much more pessimistic assumptions than have been employed in the DOE studies. James L. Plummer, "Methods for Measuring the Oil Import Reduction Premium and the Oil Stockpile Premium," *The Energy Journal* 2, no. 1 (January 1981): 1-18.

41. Glen Sweetnam, "Reducing the Costs of Oil Interruptions: The Role of the Strategic Petroleum Reserve," Presented at the ORSA/TIMS Joint National Meeting, Colorado Springs, Colorado, November 10, 1980.

42. The $20 billion figure assumes a loss of $80 billion for one million barrels per day of year-long shortfall and leakage of two-thirds of each barrel of drawdown to other nations.

43. U.S. Congress, Senate, Committee on Governmental Affairs, "Oversight of the Structure and Management of the Department of Energy," Staff Report, December 1980 (Washington, D. C.: Government Printing Office, 1980), pp. 260-61.

44. Ann Pelham, "Energy Department Trying to Work Out Problems of Costly Oil Storage Program," *Congressional Quarterly Weekly Report*, February 3, 1979, pp. 204-5.

45. Arnold Meltsner defines policy entrepreneurs as persons with enough analytical and political skills to be players in the policy process. Arnold J. Meltsner, *Policy Analysts in the Bureaucracy* (Berkeley: University of California Press, 1976), pp. 48-49.

46. This contrasts with the style of his predecessor, who more explicitly took into consideration political feasibility. Staff members who served under both assistant secretaries describe the style of the predecessor as finding politically feasible alternatives and then using the analysis of the staff to support them. In the long run, being a constant and highly visible player in the policy process may make it difficult to play an effective role in influencing decisions with analysis.

47. The associate director for Natural Resources, Energy, and Science of OMB from early 1980 to the end of the Carter Administration was Katherine Schirmer, formerly an assistant to Stuart Eizenstat, presidential advisor for domestic affairs. Secretary Duncan reportedly was hesitant to challenge her because of her strong White House connections. For a discussion of this point and a comparison of the style of the two assistant secretaries for Policy and Evaluation, see Christopher Madison, "The Energy Department at Three—Still Trying to Establish Itself," *National Journal*, November 4, 1980, pp. 1644-49.

7

The Role of the Private Sector in Strategic Stockpiling

Many major industrialized nations have established strategic petroleum reserves by requiring their refiners and importers to hold stocks of crude oil or petroleum products in excess of those held for routine operations. Even before the Arab oil embargo increased concern in the industrialized nations over the vulnerability of their economies to disruptions in the world oil market, several major importers, including France, Japan, and the Federal Republic of Germany, imposed minimum storage requirements on their petroleum industries. More recently other nations have turned to mandated industrial petroleum reserves to meet their stockpiling responsibilities as members of the International Energy Agency.

Although Congress provided discretionary authority for mandated industrial petroleum reserves in the Energy Policy and Conservation Act of 1975 (EPCA), the authority has not been implemented. Nevertheless, greater private sector involvement in the development of strategic reserves has been actively considered several times since 1975. During preparation of the SPR Plan, requirements were rejected for reasons of cost and potential implementation problems. Later, delay in the SPR Program prompted the "turnkey" initiative that attempted to involve the private sector on a voluntary basis in the construction and management of storage facilities. Recently, controversy over delays in filling available SPR capacity and the belief that adding capacity soon would provide substantial net benefits have led to reconsideration of industrial petroleum reserve (IPR) options, including the creation of quasi-public corporations and subsidy programs. A review of these proposals will explain why the United States has rejected an industrial role in the creation of strategic reserves in the past and suggest desirable ways of structuring such a role in the future.

At the outset we need to answer a fundamental question: Will private firms make socially optimal stockpiling decisions in the absence of government intervention? Welfare economics argues persuasively that, in the absence of market failures such as externalities and undefined property rights, the self-interested behavior of profit maximizing firms will lead to a socially efficient allocation of productive resources. The existence of externalities, where there is a divergence between private and social costs and benefits, implies that private actions will generally not lead to socially efficient allocations of resources. In the case of firms stockpiling in anticipation of future disruptions in the world oil market, purchases may increase the world price, inflicting external costs on domestic importers. On the other hand, sales during market disruptions may provide external benefits to domestic importers through a lower world price. Other external benefits not captured by private stockpile holders include greater flexibility in foreign affairs, deterence against purposeful market disruptions, and reduction in the macroeconomic losses caused by price shocks. The external costs lead private firms to overstockpile in the social sense; the external benefits lead private firms to understockpile in the social sense. Focusing solely on the external effects of purchases and sales, analysis employing dynamic programming suggests that in general private firms will understockpile.[1]

These external effects, however, are of secondary importance in the U.S. context. The key factor is the expectations of firms about the price they will receive when they sell stockpiled oil in the future. Risk neutral firms attempting to maximize the present value of profits will add oil to their stockpiles as long as the sum of the price of oil in the current period and the marginal cost of storage until the next period is less than the present discounted value of the expected price in the next period.[2] From 1950 to 1973 the real price of crude oil fell. It is not surprising that the industry developed a tradition of not holding speculative stocks; historical experience argued that holding inventories beyond the levels needed as working stocks in the transportation, refining, and distribution systems would not be profitable. Since 1973 there have been sharp rises in the real price of oil during disruptions in the world market that could have yielded hefty profits from the sale of oil stockpiled at previously lower real prices. If firms were allowed to sell stockpiled oil at the world price no matter what the state of the market, we would expect firms to hold at least some speculative stocks in anticipation of higher prices during future supply disruptions. Beginning with general wage and price controls in 1971, however, U.S. policies have discouraged speculative stockpiling by reducing the expected price firms believe they will be able to realize from the sale of stockpiled oil in the future. Even with the elimination of price controls on crude oil and petroleum products, firms will include in their calculations the probability that controls will be reinstituted during market disruptions. Existing regulations allow the secretary of energy to force stockpilers to sell oil at controlled prices to firms that have not stockpiled.[3] In light of these policies, it is no wonder that firms have been reluctant to accumulate voluntarily speculative stocks.[4]

The Industrial Option in the SPR Plan

The authority provided by EPCA for creation of an industrial petroleum reserve (IPR) was previewed in Senator Jackson's Petroleum Reserves and Import Policy Act of 1973.[5] Jackson's bill would have required the establishment of a reserve of sufficient size to replace all imports for ninety days through a combination of mandatory industrial storage, unutilized capacity for producing oil and natural gas from private and public lands, and stockpiles directly accumulated by the government. The particular combination of these three approaches to be employed was to be determined by an administrative committee chaired by the secretary of the Department of Interior.

Although the legislative history of the IPR component of the EPCA is sparse, several factors motivated its inclusion.[6] It would distribute a portion of the burden of the SPR to the producers and consumers of petroleum products so that the entire burden would not fall on the general taxpayer. By taking advantage of the special expertise and capabilities of the petroleum industry, it offered the possibility for more rapid development of the SPR. Additionally, the IPR approach was by then being followed by most of the other major importing nations.

As EPCA began taking shape, the small team of analysts in FEA working on strategic petroleum reserve legislation was dubious about the desirability of mandating industrial reserves. Because of Senator Jackson's strong preference for industrial reserves, however, the FEA administrator, with the approval of President Ford, decided that it would be politically advisable to allow discretionary authority for an IPR to be included in the legislation. During the EPCA debate in July 1975, Senator Jackson extracted a written promise from the FEA administrator that the IPR option would be seriously considered.

Under EPCA the administrator of the FEA (now the secretary of energy) has discretionary authority to include an IPR as part of the SPR plan:

> The Administator may require each importer of petroleum products and each refiner to (1) acquire, and (2) store and maintain in readily available inventories, petroleum products in amounts determined by the Administrator, except that the Administrator may not require any such importer or refiner to store such petroleum products in an amount greater than 3 percent of the amount imported or refined by such person, as the case may be, during the previous calendar year.[7]

If the administrator decides to implement the IPR program, EPCA requires him to do so in a manner that avoids inequitable economic impacts on refiners and importers. The administrator is permitted to grant exemptions from storage regulations to firms that "would otherwise incur special hardships, inequity, or unfair distributions of burdens." EPCA also authorized the administrator to allow selected firms to use surplus storage capacity owned by the government to meet their stockpiling requirements.

The newly created SPR Office relied upon two major information sources in preparing the IPR section of the Strategic Petroleum Reserve Plan. In January 1976, shortly after passage of EPCA, the FEA commissioned a major study of the feasibility of implementing an IPR.[8] Submitted in the summer of 1976, the draft study suggested a number of possible implementation scenarios, warned of some potential problems, and identified major uncertainties. By raising more questions than it answered, it implicitly showed the administrative complexity of any IPR program that attempts to achieve an equitable distribution of burdens. Nevertheless, it did not explicitly recommend that the IPR option not be implemented.

The second major information source indicated the hostile political environment likely to be encountered during implementation and refocused attention on the question of whether or not there should be an IPR at all. In June 1976 the FEA announced in the *Federal Register* that a hearing would be held the following month on the feasibility of implementing the IPR.[9] Most of the respondents were petroleum firms and trade associations opposed to implementation. Of the forty-four organizations that submitted written comments or offered testimony, only one favored IPR implementation.

Opponents argued that storage requirements would inflict heavy capital costs on the industry, which would divert investment from exploration, production, and refinery modernization. Many argued that the diversity of the petroleum industry in terms of size, financial structure, and type of business would necessarily result in an unequal ability to meet the capital costs. Others argued that attempts to equalize burdens through exemptions to particular firms would leave other firms at a competitive disadvantage.

Several of the organizations suggested that the IPR would be challenged on constitutional grounds. Depending on the level of control the government maintained over the mandated reserves, the courts might rule that the IPR violated the Fifth Amendment guarantee against the taking of property without compensation. Even if legal challenges were ultimately resolved in favor of the government, the implementation of the IPR could be seriously delayed because the IPR was to be a component of, rather than a supplement to, the total SPR storage; it was believed at the time that an attempt to implement the IPR in the face of legal challenge might prevent realization of the goals established by EPCA.

The Strategic Petroleum Reserve Plan, completed by the SPR Office in December 1976, recommended that the authority for creation of the IPR not be exercised. The chapter dealing with the IPR presented six primary reasons for the decision:[10]

An IPR would not accelerate the development of the SPR. The SPR Office believed that there was little excess storage readily available to industry for IPR use. If the IPR was going to constitute additions to reserves above existing operating stocks, firms would have to build new storage or use storage constructed by government. It was assumed that firms would not be able to build storage capacity more quickly than the government. Although a plausible assumption at the time, subsequent experience has shown it to have been based on an overly optimistic assessment of how quickly the government could develop salt dome storage capacity.

Any Regional protection that might be provided by an IPR could be achieved more efficiently and effectively with a government-owned reserve. EPCA seems to require the storage of products in any U.S. region that imports more than 20 percent of its refined products. However, it permits the storage of crude oil in a nearby region as an alternative to storage of products directly in the region. The SPR Plan took advantage of this loophole in rejecting the initiation of regional storage of products. Because of strong political pressure (primarily from senators and congressmen representing states in New England), persons both inside and outside of the SPR Office considered the IPR as a potential way of achieving regionally distributed product reserves without the construction of government facilities. In rejecting this approach, the SPR Plan argued that, to the extent it resulted in regional storage, the IPR would involve higher cost storage and reduce the flexibility of the SPR by placing reserves in locations from which they could not be readily moved to areas with greater demand during supply disruptions.

An IPR is likely to result in higher costs to the national economy as a whole. It was assumed that much of the IPR storage capacity would be in the form of newly constructed steel-tank storage. The SPR Office estimated that the capital costs of this storage would be five to ten dollars per barrel greater than the salt dome storage planned for the SPR.

An IPR may delay the SPR program because of legal challenge, and it could create substantial programmatic and environmental problems. In addition to the issue of the taking of property without compensation, it was unclear if firms could be required to store crude and product slates different from their imports and product outputs. If firms stored a percentage volume of each type of crude they imported and product they made, a large portion of the IPR would consist of crudes and products of little value during disruptions. The administrative task would be highly complex because firms routinely alter their input and output slates. The decentralized storage would also involve greater environmental risks than centralized salt dome storage; firms might face delays in capacity development due to problems encountered in the environmental permitting process.

An IPR could result in adverse impacts on the competitive environment within the petroleum industry and upon the competitive position of individual firms. The SPR Plan noted that larger firms were more likely to be able to exploit economies of scale in the construction of storage facilities. (It did not discuss the possibility that firms, using efficient large-scale facilities, might arise to handle the reserve requirements of smaller firms on a fee basis.) At the time, recent experience with petroleum industry regulatory programs, such as the small refiner bias in the Entitlements Program and the forced allocation of oil to small refiners inherent in the Oil Import and Crude Oil Buy-Sell Program, suggested that political forces would encourage excessive modification of rules in favor of small refiners. Regulatory complexity of the IPR would steadily grow as attempts were made to relieve special hardships and better balance burdens.

The shifting of costs from the U.S. Government to the petroleum industry (and to consumers of petroleum products) is the only apparent advantage of an IPR, but this does not in itself offer significant economic or conservation benefits. The SPR Plan estimated that the IPR could save as much as $4 billion in federal budget outlays over an eight-year period. These expenditures, however, would be shifted to industry so that reductions in government spending would not leave more resources for the private sector. In fact, the regulatory approach would increase government interference in private affairs. By increasing the costs of the petroleum industry, the IPR would result in slightly higher prices for consumers and hence encourage conservation. However, the SPR purchases would increase gasoline prices only one-tenth to two-tenths of a cent per gallon. Such a small price rise would have a negligible impact on demand.

Did the SPR Office neglect any important considerations in rejecting the IPR option? In retrospect, two factors make an IPR begun in 1977 appear somewhat more attractive than analysis presented in the SPR Plan indicated.

First, the IPR could have been begun in conjunction with development of the centralized salt dome storage capacity for Phases I and II (500 million barrels). If the IPR had proved to be administratively feasible, part of the Phase II construction could have been cancelled with only moderate loss of wasted effort. If the IPR had proved to be administratively unfeasible, it could have been cancelled. The parallel approach would have traded increased cost for a greater chance of having storage capacity available on schedule. Of course, it might be argued that proceeding with the IPR would have further overburdened the SPR Office staff and thus contributed to even greater delay than actually was encountered in the implementation of Phases I and II. This diversion of managerial resources might have been avoided by transferring administrative authority for the IPR to the part of FEA (now the Economic Regulatory Administration) that was already involved in the regulation of the petroleum industry.

Second, the SPR Plan did not anticipate that the U.S. Government would face threats of retaliation from producers for filling the SPR. When the SPR Plan was written, the world oil market was slack and was expected by many petroleum economists to remain slack into the early 1980s. Under slack market conditions, it would be difficult for any single producer to retaliate by reducing production because unused production capacity existed elsewhere. The situation changed dramatically with the loss of Iranian production in 1979. The U.S. Government delayed resuming SPR purchases in the wake of the Iranian revolution because of Saudi Arabian threats even after prices stabilized. The U.S. experience contrasts with the experiences of Japan and several Western European nations that have continued to build reserves through IPR programs without being threatened by producer retaliation. The low visibility of purchases for reserves by private firms suggest that an IPR for the United States would also have escaped threats of retaliation.

If the IPR had been started in 1977, there is a possibility that we would now have an additional 185 million barrels of oil in storage—almost double the amount in storage in the SPR at the end of 1980. Nevertheless, even considering the factors neglected by the SPR Office, it would be difficult to argue that the decision to implement the IPR was incorrect at the time; in comparison to the IPR, the direct development of reserves by the government was politically less controversial and appeared to involve fewer implementation problems. As discussed in the next two sections, the parallel development and acquisition factors prompted later reconsideration of an industrial role in the development of strategic reserves.

The Turnkey Initiative

When retired General Joseph DeLuca became the director of the SPR Office in July 1978, he assumed responsibility for a program very much behind schedule and over budget. He devoted most of his effort to improving the management and integration of the 528 million barrels of capacity under development at sites already selected by the SPR Office. For development of the next 222 million barrels of capacity (Phase III), he proposed contracting with private firms on a fixed price basis for completed facilities rather than for components of facilities. This approach became known as the "turnkey" program because firms would not be paid until they turned over the keys for fully operational facilities to the government. It is similar to the fixed price contract sometimes employed by the defense department.

General DeLuca viewed the turnkey program as having several advantages over the direct development of Phase III by the SPR Office.[11] First, the turnkey program would greatly streamline the procurement process. A fragmented procurement system involving hundreds of government contracts for materials, equipment, and construction at each site had contributed to the failure of the SPR program to meet its schedule and cost goals. The turnkey program would employ a single fixed-price contract for each site. Instead of having to monitor the work of numerous subcontractors, the government only would have to determine if the completed facility met the established specifications. The turnkey program would essentially take the SPR Office out of the general contracting business.

Second, the turnkey program would allow the SPR Office to secure additional sites without resort to the powers of eminent domain. The seizure of land through condemnation caused considerable animosity toward the SPR at the local level, particularily in Louisiana, where Governor Edwin Edwards was generally critical of the program. Under the turnkey program, private firms would assemble land for sites through voluntary market transactions as they normally would for any industrial project.

Third, the turnkey program would force industry to put its "expertise on the line." General DeLuca reports that he was regularily visited by the presidents of firms who argued that they could do a better job in providing storage capacity. Criticism from firms that did not offer turnkey bids would be more easily deflected. From the bids that were offered, a market standard for SPR program performance would be provided. Additionally, it was hoped that the turnkey projects would suggest innovative technical approaches.

Fourth, the turnkey program would force OMB to take an explicit position on Phase III. OMB had failed in its attempt to stop the Carter administration from making a commitment to an expansion of the SPR to one billion barrels (Phases III and IV) by the end of 1985. When problems with the implementation of the first two phases became apparent, OMB argued that Phase III should be delayed until the SPR Office was better prepared to manage its implementation. This argument would not be appropriate for delaying turnkey projects because they did not depend heavily on SPR Office management capabilities. Consequently, the turnkey program would force OMB to attack expansion beyond Phase II capacity directly rather than through delay.

Although his SPR Office staff was generally opposed to initiation of the turnkey program, General DeLuca decided to proceed after receiving backing from Secretary Schlesinger.[12] A source evaluation board (SEB), chaired by James R. Higgins of the Procurement Office, was established to solicit and evaluate turnkey proposals. Except for technical assistance provided by several members of the SPR Office staff, the turnkey SEB operated as an independent organization. At the same time, the SPR Office arranged for the U.S. Army Corps of Engineers to serve as the lead agency in managing the compliance of turnkey projects with the requirements of the National Environmental Policy Act.[13] These delegations of responsibility relegated the SPR Office to a minor role in the development of Phase III facilities through the turnkey program.

The turnkey program consisted of two separate initiatives: noncompetitive negotiations with owners of sites with existing storage capacities and competitive procurement of additional sites. The noncompetitive procurement was justified as necessary to provide storage capacity in the interim between the completion of Phase I and the beginning of availability of capacity from Phase II. Negotiations were conducted with the owners of three facilities. Sites located in Louisiana at Cote Blanche and Napleonville would have each provided 30 million barrels of capacity. A third site at Irontown, Ohio, would have provided 20 million barrels of capacity. Although negotiations continued after cancellation of the competitive turnkey solicitation in August 1979, by early 1980 all negotiations were cancelled or indefinitely suspended.

The competitive turnkey solicitation was intended to lead to the procurement of at least 142 million barrels of capacity by the end of 1985. At a presolicitation conference in October 1978, General DeLuca issued a challenge to industry to provide as much as 600 million barrels of capacity by the same date. If industry had met this challenge, sufficient capacity would have been available to reach the goal of one billion barrels of oil in storage by the end of 1985 as established by Amendment Number 2 of the SPR Plan.

In the first phase of the turnkey solicitation, issued in November 1978 to 160 interested parties, firms were asked to focus primarily on technical aspects of facility development. It was indicated that proposals would be evaluated in terms of four criteria: compatibility with overall national supply and distribution system, compatibility of proposed schedule with the SPR Plan, storage capacity (minimum: 20 million barrels), and risk of failure to complete facility on schedule. Firms were given the option of offering filled or unfilled facilities. By the January 1979 deadline twenty-three proposals were submitted.

The SEB spent several months discussing the proposals in detail with the firms. By April 1979 eleven proposals were either rejected or withdrawn. The remaining twelve were considered sufficiently attractive to be included in the second phase of the turnkey solicitation, the negotiation of fixed price contracts. Of the eleven firms that participated, none were willing to offer bids for filled facilities because of the great uncertainty about future oil prices.

Between June and August 1979, the SEB conducted three rounds of price negotiations.[14] Initially, firms were asked to submit a per barrel bid for a twenty-year period with equal payments in each year beginning when the facility was completed. That is, a twenty dollar per barrel bid would result in payments of one dollar for each of the twenty years after the facility was opened. In fact, the initial bids were only slightly less than twenty dollars per barrel. At the time the highest cost SPR facility was estimated to have construction costs averaging $3.85 per barrel of capacity upon completion.[15]

Several factors help explain the apparently high turnkey costs. The extended payment period and the risk of liability for large damages in the event of oil leakages, fires, or spills made it difficult for firms to secure financing for construction. The delay of payment until after completion of facilities further worsened the financing picture. Comparing the turnkey and SPR program costs in terms of present value greatly reduces their disparity. The government must make all the SPR expenditures before the facility is open so that the undiscounted costs are only slightly higher than their present value. The present value of the turnkey costs, on the other hand, would be much smaller than their undiscounted sum because they would be paid by the government over a much longer period. For example, at a discount rate of 10 percent, the present value of government payments of one dollar for each of four years is $3.49, as compared to a sum of undiscounted costs of $4.00. The present value of payments of one dollar per year for twenty years beginning five years in the future is $6.40, as compared to a sum of undiscounted costs of $20.00.

In the second round the SEB asked firms to submit bids under the assumption that they would receive government loan guarantees. In the third round firms were asked to submit bids under the assumptions of loan guarantees and yearly payments beginning when contracts were signed rather than when facilities were completed. The third round undiscounted bids were about double estimated costs for the SPR Office development of new capacity. When discounted, however, the costs were much closer. A fourth round of bids assuming full government payment (for purchase or twenty-year lease) at completion of facilities would have greatly reduced the problem firms faced in anticipating interest rates and likely have led to bids very close to the estimated costs of direct government development of new capacity. In effect, the fourth round would have constituted the negotiation of fixed-price performance contracts.

As it happened, negotiations were terminated during the third round. DOE announced cancellation of the turnkey solicitation on August 30, 1979. Shortly after, DOE Under Secretary John Deutch explained the rationale for the decision in a prepared statement to the House Subcommittee on Energy and Power:

> This decision was based on a change in SPR Program requirements. The solicitation called for the earliest possible fill with oil availability not a constraint. Also, the turnkey solicitation called for a long-term (20-year) commitment by the Federal Government to special purpose facilities. The Iranian crisis, recent ceilings on imports, and the tight crude oil market are indicative of the serious constraint which crude availability has become to the SPR Program. In light of this uncertainty, it

> would not be in the best interest of the Government to enter into such long-term commitment, based on fill schedules and other requirements specified at the present time. In addition, the turnkey proposals demonstrated a limited industrial acceptance of liability and responsibilities for storage containers and crude oil with no apparent cost advantages or technical innovations to offset the programmatic risks.
>
> In view of the above factors the need to maintain flexibility in the development of the remaining necessary SPR facilities, a different facilities development approach appears to be needed.[16]

The logic of basing storage capacity development decisions on current market conditions is questionable. The lead time for turnkey projects is four or five years; it is even longer for capacity additions by the SPR Office. During such a long period, the market is likely to alternate several times between periods of tightness and slack. In fact, periods of slack are likely to follow major disruptions, as supply that had been temporarily removed from the market is returned. The market was generally slack for several years following the end of the Arab embargo in 1974. Only nine months after Under Secretary Deutch made his statement, the world market was again slack with record involuntary increases in private inventories and the shutting-in of production capacity by several producers to maintain prices. The war between Iraq and Iran tightened the market in late 1980, but by the spring of 1981 the market was again slack with the beginning of a decline in nominal prices.

The decision may have reflected a reassessment within the Carter administration of the commitment to expand the reserve to one billion barrels. Under pressure from OMB, DOE agreed to conduct a new analysis of the economic costs and benefits of Phase III and IV to take into account higher oil prices and greater facility development costs. At the highest levels of the administration, balancing the budget may have risen in relative importance to meeting SPR goals. The turnkey projects would have involved greater near-term budget costs than the slower paced development of salt dome capacity by the SPR Office. The Iranian crisis provided a public rationale for delaying SPR expansion.

The decision may also have reflected a reluctance on the part of DOE to seek from Congress the approval of special provisions needed for the completion of turnkey negotiations. If the turnkey facilities were going to be leased at reasonable cost, some sort of loan guarantees and risk sharing provisions needing congressional approval would have to be provided. If the turnkey facilities were to be purchased upon completion instead, large budget commitments would be required. Securing congressional cooperation would have been difficult. The credibility of DOE with respect to the SPR program was at a low point following the disclosures of cost overruns and failures to meet schedules during the previous year. Congressman John Dingell, probably the most influential congressional advocate for completing the SPR program, had expressed considerable skepticism about the turnkey initiative in a hearing held shortly after the turnkey solicitation was begun.[17] With the prior departure of General DeLuca, there was no longer a strong advocate for the turnkey approach in the SPR Office anxious to invest the effort needed to secure congressional approval.

The turnkey solicitation was a lost opportunity for expeditiously adding Phase III (and perhaps Phase IV) facilities. It is conceivable that the turnkey approach, by providing the parallel development of a number of facilities, could have been used to provide sufficient storage capacity to meet the one billion barrels by the end of 1985 goal set out in the second amendment to the SPR Plan. Although budgetary costs most likely would have been greater than those for direct government development of facilities, benefits associated with the opportunity for earlier fill, such as making lower cost purchases during slack markets and having greater drawdown capabilities sooner, could very well have more than offset the higher costs.[18] In fact, analysis suggesting that large net benefits are associated with earlier development has prompted yet another round of consideration of the industrial role in the SPR.

Other Proposals For Private Sector Involvement

In 1980 several factors motivated consideration of nongovernmental reserve options by analysts in the DOE Office of Policy and Evaluation (PE) and by several key SPR advocates in Congress. First, the reluctance of the Carter administration to resume filling the SPR for fear of offending Saudi Arabia (and perhaps for budgetary reasons as well) drew attention to the political factors that interfere with economically efficient acquisition decisions. While the government refrained from making purchases, private firms were involuntarily accumulating record levels of crude oil inventories that they would have been happy to sell to the government for the SPR. This phenomenon, along with the observation that foreign IPR programs built reserves without threats of retaliation by exporting nations, suggested that individual firms or an independent stockpiling corporation would be better able to accumulate oil for strategic reserves.

Second, analyses by PE of optimal acquisition and drawdown policies for the SPR indicated that accumulating larger reserves sooner would produce large expected net social benefits under the most plausible sets of assumptions.[19] Because it was committed to salt dome leaching at a limited number of sites, the SPR program could not be easily accelerated. Expanding private sector stockpiling, either directly through regulation or indirectly through subsidies, was viewed as a way of accelerating the accumulation of total reserves. Additionally, incentives for encouraging private stockpiling might be structured to permit the implementation of more nearly optimal drawdown strategies. Analysis suggests that in many circumstances it is advantageous to drawdown reserves for minor market disruptions. It is unlikely that the reserves directly controlled by government would be released for minor disruptions. Private firms voluntarily holding reserves in response to incentive programs, however, might find it profitable to reduce inventories when minor disruptions cause more moderate price rises. A mixed system of government and expanded private reserves would be more flexible than the governmentally controlled SPR alone.

Third, removing costs from the federal budget was seen as a way of reducing the opposition of OMB to SPR expansion. Even though an IPR program would involve higher economic costs, it would reduce budget outlays. In fact, in 1978 OMB requested that DOE reconsider the IPR option as an alternative to expansion of the SPR beyond 500 million barrels. The budget argument was also relevant to the fill question. If the SPR were run by an independent corporation with its own budget authority, the long-term benefits of purchasing reserves would no longer suffer from the exigencies of the federal budgetary process.

During the summer of 1980, a major study of policy options for reducing U.S. vulnerability to oil supply disruptions was conducted under the direction of William Lewis, then assistant secretary for

Policy and Evaluation. The final report called for consideration of three IPR options: the implementation of mandatory storage regulations under existing or modified EPCA authority, the provision of subsidies and tax incentives to induce private firms to hold larger reserves, and the creation of an IPR corporation to facilitate the holding of reserves by private firms.[20] The financial incentives and IPR corporation were advanced as ways of mitigating the problems likely to be encountered in implementing a mandatory industrial program.

The West German experience suggests how these options might work in practice.[21] In 1975 West Germany increased storage requirements for refiners and importers affiliated with foreign refiners and initiated storage requirements for the previously exempt independent importers. The government offered aid in the form of loan guarantees and tax exemptions to help reduce the costs to firms holding mandated stocks.[22] Nevertheless, both independent and dependent importers challenged the new requirements in the German Federal Constitutional Court. Before the challenges were resolved, an agreement was reached among the various components of the petroleum industry and the government to create a public corporation that would hold required stocks for member firms for a fee proportional to the amount of each type product refined or imported. The corporation, called the Erdoelbevorratungsverband (EBV), borrowed sufficient funds to secure storage facilities and accumulate mandated levels of stocks upon its creation in the autumn of 1978. Interest on loans is paid through product fees and equity is to be built up when stocks are sold at market prices during drawdowns authorized by the government.

The EBV enables the German petroleum industry to take advantage of economies of scale in meeting storage requirements. More importantly, the uniform fees it levies do not alter the competitive position of the firms; the fees can be viewed simply as excise taxes paid by all firms in the industry. Because the EBV holds all the mandated stocks, individual firms do not have to use their borrowing ability to finance their reserve requirements. As an additional benefit, the EBV provides a clear distinction between crude oil and products that are part of the mandated reserves and those held by firms as working stocks.

As was the case in West Germany, tax subsidies probably would be an insufficient palliative to overcome the opposition of U.S. firms to the imposition of mandatory reserve requirements. The tax subsidies would not necessarily reduce the inequities that would result from uniform storage requirements. A tax subsidy program with voluntary participation would suffer from the "maintainance of effort" problem; it could very well subsidize reserves that would have been held anyway. From a political standpoint, tax subsidies would suffer because they would appear to be giving the oil industry benefits at a time when many of the firms are showing record profits. As long as firms anticipate the possibility of the reimposition of price controls during supply disruptions, the level of tax subsidies that would be politically feasible most likely would be too small to induce substantial increases in private sector stockpiling.

Perhaps the subsidy approach would be most effective for encouraging electric utilities to build reserves of residual fuel oil. Because state public utility commissions often do not allow an immediate pass-through of higher fuel costs by utilities to their customers, the possibility that price controls on petroleum products would *not* be imposed during disruptions should provide some incentive for stockpiling. Current DOE regulations, however, make it illegal for utilities to build stocks beyond normal levels. Repeal of these regulations and providing direct or tax subsidies to reduce the costs of storage facilities and the interest cost of holding stocks could lead to a significant increase in the levels of reserves held by these users. A disadvantage of this approach is that it involves the higher cost of storing residual fuel oil which is of less economic value during disruptions than equal volumes of crude oil.

Instituting an IPR through an EBV-like corporation is more promising but might also be difficult to implement in the U.S. political context. The EBV can be thought of as a corporation that has authority to collect "taxes" (fees) on petroleum products in order to pay the storage and interest costs of meeting a mandated industry requirement. Recent proposals to increase taxes on products, such as President Carter's attempt to raise gasoline taxes by ten cents per gallon, have been blocked by fierce political opposition. An EBV-like corporation for the United States might likewise be vigorously opposed if it came to be viewed primarily as a mechanism for imposing excise taxes on petroleum products. It would be opposed by the petroleum industry unless the imposition of an IPR program was considered certain.

If an EBV-like corporation was created, it would only provide competitive neutrality if all firms in the petroleum industry were forced to become members. Under voluntary membership, firms with storage costs lower than the average for the corporation would elect to meet their storage obligations on their own. As a result, they would gain a competitive advantage over the firms whose lowest cost option was membership in the corporation. Nevertheless, the resulting distribution of costs would be fairer than under an IPR alone. In fact, if an IPR was implemented, we would expect private companies to arise to perform the centralized storage function of the EBV for firms facing high storage costs because of their size, location, or product mix.

Several proposals have been advanced for operating the SPR less as a government agency and more as an independent entity that interacts to a greater extent with the private sector. They represent an attempt to substitute market decisions by firms and individuals for political decisions concerning oil acquisition and, in some proposals, drawdown.

The accumulation of stocks by industry in high cost tanker storage coincident with the reluctance of the Carter administration to resume purchases for the SPR prompted proposals that would enable firms to transfer unwanted stocks to the government. Former acting SPR Director Carlyle Hystad suggested that the SPR be temporarily made available for storage of oil by firms.[23] Under his plan, firms would have been charged a minimal fee for storing their oil in the SPR. After a fixed period of time, firms would have been allowed to withdraw their oil with the government exercising the right to purchase it.

A more general plan for opening up the SPR for private storage had been previously suggested by Joseph Nye of Harvard University. The plan called for the government to increase or decrease the amount of oil in the SPR by adjusting the fee charged to private firms for storage. To accumulate reserves the government would offer storage space at a low, perhaps negative, fee. To induce withdrawals during disruptions, the government would increase the storage fee (or reduce the storage subsidy) until the desired rate of drawdown was achieved. A major advantage of this plan is that the government would not have to purchase oil directly in the world market and hence would be less susceptible to threats from producing nations. The plan would also increase the chances that the SPR reserves would actually be used during disruptions. Whereas the government may be reluctant to reduce reserves during minor supply disruptions out of fear of having less than full reserves in the event of a subsequent major disruption, some firms probably would heavily discount future disruptions and therefore would reduce their reserves to take advantage of moderate price increases. Instead of the government having to make highly visible decisions about acquisitions and drawdowns, a continuous adjustment of the size of the reserve, reflecting the aggregate expectations of the petroleum industry, would result. Additionally, it would permit firms most vulnerable to supply disruptions to diversify against risk by taking advantage of the low-cost SPR storage. The major disadvantage of the plan is that it might require politically unpopular subsidies to be given to industry. As long as firms believed the nominal price of oil would not decline, the subsidy rate needed to induce storage in the SPR probably would not be greater than the rate of interest the firms faced. Nevertheless, providing any level of subsidy, including charging a fee less than the cost to the government of providing the storage, would be controversial.

As discussed in chapter 4, an independent SPR corporation could be financed by the sale of certificates having redemption values tied to the price of oil. Such a system of equity financing would allow persons and firms to share in the costs and benefits of the SPR program. In particular, it would give those who expect to suffer most from an oil supply disruption an opportunity to reduce their risk through the purchase of an asset (the certificate) that will appreciate in value during disruptions. Although equity financing was rejected during the 1981 debate over moving SPR oil purchases off-budget, it is likely to be reconsidered in the future.

Proposals for private sector involvement in strategic stockpiling will continue to be offered as alternatives to expansion of the slow paced SPR program. The final one we will consider here was suggested in the summer of 1981 as a way of converting excess private sector stocks into strategic reserves. Although industry stocks of crude oil were then at record levels, firms were beginning to reduce their stocks in the face of high interest rates (which impose high carrying costs), reduced demand (which reduces the need for seasonal stocks), high physical storage costs (marginal storage in tankers), and expectations of falling crude prices (which reduces the desire to hold speculative stocks).[24] Analysts in the Office of Energy Security in DOE proposed converting some of the excess stocks to strategic reserves through the auctioning of a type of sell-option. Firms willing to hold oil continuously in storage for a fixed number of years would bid for the right to sell the oil to the government at the end of that time for a predetermined price. The price would be set sufficiently high to ensure at least some bids.

For example, assume that the government offered five-year sell-options of $135 per barrel at the beginning of 1983. That is, firms that entered into contracts would store oil from the beginning of 1983 to the end of 1987, at which time they would receive $135 per barrel. Now consider a firm that faces a 20 percent annual interest rate and employs a 20 percent annual discount rate in computing the present value of investments. If oil costs $35 per barrel, the carrying costs for each barrel of oil would be $7 per barrel per year in interest and $4 per barrel per year in physical storage costs (a figure slightly higher than current tanker rental charges), the present value of these costs would be $35.88. The present value of the per barrel return at the end of 1987 ($135 less $35 as repayment of principal) would be $48.20 per barrel. For the opportunity to enter into the contract, the firm would be willing to bid up $12.30 to maximize its profits. A firm with lower storage or capital costs should be willing to bid even more.

The sell-option proposal has a number of desirable characteristics. It would lock in stocks that would otherwise be drawn down, increasing strategic reserves in the short run. It would require little information to implement, as long as the amount under contract is limited so that the bidding will be competitive and excess profits eliminated. When sell-options come due, the oil could be transferred to the SPR as storage capacity becomes available. Finally, it would be relatively easy to enforce; firms found not maintaining the sell-option stocks as separate reserves would simply lose the option.

However, the sell-option proposal would be vulnerable to charges that it benefits the budget of the current administration to the detriment of budgets of future administrations. It is also likely that questions would be raised about the competitiveness of the bidding and hence the propriety of high sell-option prices. Overall, the sell-option proposal is probably one of those ideas that seems only natural to economists but outrageous to politicians and the public.

Notes

1. Thomas J. Teisberg, "A Dynamic Programming Model of the U.S. Strategic Petroleum Reserve," *Bell Journal of Economics* 12, no. 2 (Autumn 1981):526–46.

2. Mathematically, the risk-neutral, profit maximizing firm will increase its stockpiles as long as

$$(P_O + v) < E(P_1)/(1 + r)$$

where P_O is the current price, v is the cost of holding a unit of oil until the next period, $E(P_1)$ is the expected price in the next period, and r is the rate of interest.

If there is a large probability of stable prices and a corresponding small probability of very high prices, speculation will appear very risky: a large probability of net costs and a small probability of large net profits. Risk averse firms would therefore stockpile less than risk neutral firms.

3. The Code of Federal Regulations states:

(a) Inventories of crude oil and allocated products. No refiner, importer, wholesale purchaser or end-user shall accumulate inventories of any crude oil or allocated product which exceed customary inventories maintained by that refiner, importer, wholesale purchaser or end-user in the conduct of its normal business practices unless otherwise directed by the DOE. Normal inventory practices shall be observed in determining allocable supplies of crude oil or allocated products in each period which corresponds to a base period. The DOE may review inventory practices and direct and increase or decrease inventories if:

1. the inventory practices employed are inconsistent with the provisions of this part;
2. the inventory practices circumvent or otherwise violate other provisions of this part; or
3. the DOE determines that an adjustment is necessary in order to allocate crude oil or allocated product supplies consistent with the objectives of the Mandatory Petroleum Allocation Program. (10 CFR 211.22, Revised as of January 1, 1980.)

Charles Phelps, who brought this regulation to my attention, has raised an interesting constitutional question: Is it possible for Congress to provide a guarantee to stockpilers that their right to sell at market prices will not be restricted by a future Congress? Without such a guarantee, we should not expect private firms to stockpile at a socially optimal level.

4. During the summer of 1980, private inventories of crude oil greatly increased. Consumption had fallen dramatically in response to price increases of the previous year and a general slowdown in the world economy. At the same time firms continued purchasing oil from nations at high rates so as not to jeopardize future lifting rights. (Some exporters, such as Venezuela and Nigeria, could not find buyers at current prices and began shutting in productive capacity.) As a result, inventories expanded greatly, so that in early September an estimated 200 million barrels above normal were being held in tankers at sea. These fortuitously accumulated stocks averted sharp price rises at the beginning of the war between Iraq and Iran.

5. Petroleum Reserves and Import Policy Act, S. 158, 93rd Cong., 1st sess., April 16, 1973.

6. JRB Associates, Inc., "Feasibility Study for Requiring Storage of Crude Oil, Residual Fuel Oil and/or Refined Petroleum Products by Industry," Final Report to the Federal Energy Administration, December 2, 1976, p. I-2.

7. Energy Policy and Conservation Act, Public Law 94-163, Title I, Part B, Section 156.

8. JRB Associates, Inc., "Feasibility Study."

9. Ibid.

10. Strategic Petroleum Reserve Office, "Strategic Petroleum Reserve Plan," December 15, 1976.

11. Interview, December 16, 1980.

12. General DeLuca's perception of staff opposition is consistent with reports by Carlyle Hystad that he and others opposed the turnkey concept because earlier contacts with industry suggested that their bids would not be competitive with government development. Interview, September 17, 1980.

13. U.S. Army Corps of Engineers, "Management Plan Development of Environmental Impact Statements for Turnkey Sites," Huntsville, Alabama, August 20, 1979.

14. My discussion of the second phase of the turnkey solicitation is based on public documents and interviews with several persons directly or indirectly involved in the initiative. Although records of the solicitations still exist and would be an excellent source of data about the probable "supply curve" of privately provided storage capacity, DOE has decided to withhold the distribution of such data, even for internal use. Several of the firms are suing DOE for the costs they incurred in preparing proposals on the ground that DOE entered into negotiations in bad faith. Use of the data might support the argument that DOE began the turnkey initiative solely to gather data with no intention of actually awarding contracts.

15. Strategic Petroleum Reserve Office, "Annual Report," DOE/RA-0047, February 16, 1980, p. 28.

16. U.S. Congress, House, "Strategic Petroleum Reserves: Oil Supply and Construction Problems," Hearing before the Subcommittee on Energy and Power of the Committee on Interstate and Foreign Commerce, 96th Cong., 2d sess., September 10, 1979, p. 26.

17. U.S. Cong., House, "Strategic Petroleum Reserve: Reprogramming of Funds," Hearing before the Subcommittee on Energy and Power of the Committee on Interstate and Foreign Commerce, 95th Cong., 2d sess., December 18, 1978, pp. 3-4.

18. For an explanation of this tradeoff see Glen Sweetnam, "Reducing the Costs of Oil Interruptions: The Role of the Strategic Petroleum Reserve," Presented at the ORSA/TIMS Joint National Meeting, Colorado Springs, Colorado, November 10, 1980.

19. Glen Sweetnam, George Horwich, and Steve Minihan, "An Analysis of Aquisition and Drawdown Strategies for the Strategic Petroleum Reserve," Draft, U.S. Department of Energy, Assistant Secretary for Policy and Evaluation, Office of Oil Policy, December 17, 1979; Jerry Blankenship, Mike Barron, Joseph Eschback, Linsay Bower, and William Lane, "The Energy Problem: Costs and Policy Options," U.S. Department of Energy, Assistant Secretary for Policy and Evaluation, Office of Oil Policy, May 23, 1980.

20. U.S. Department of Energy, Assistant Secretary for Policy and Evaluation, "Reducing U.S. Oil Vulnerability: Energy Policy for the 1980s," November 10, 1980. (Washington, D. C.: Government Printing Office, 1980), pp. II-A-8, II-A-9. These options were also considered in David A. Deese and Joseph S. Nye, eds., *Energy and Security* (Cambridge, Mass.: Ballinger Publishing Company, 1981).

21. For a more detailed discussion see Edward N. Krapels, *Oil Crisis Management: Strategic Stockpiling for International Security* (Baltimore: Johns Hopkins University Press, 1980), pp. 65-68.

22. Japan also provides subsidies to its industry in the form of loan guarantees, accelerated depreciation, and lower tax rates on storage facilities.

23. Sobotka and Company, Inc., "Option for Placing Current Excess Private Crude Oil Stocks in the SPR," draft, August 29, 1980.

24. Bob Jippe and Richard Wheatley, "High Cost of Bulging Inventories Compounds Problems for U.S. Refiners," *Oil And Gas Journal*, June 22, 1981, pp. 19-22.

8

Distribution and Drawdown

Under what circumstances should oil stored in the SPR be withdrawn and distributed? Explicit drawdown policies have been slow to evolve. It is not surprising that issues related to the development of storage capacity and the acquisition of oil have so far received greater attention from analysts and decision makers than the formulation of drawdown policies: until there is a significant volume of oil in storage, resolving drawdown issues does not seem urgent. The absence of explicit policies, however, is not due totally to neglect. It reflects the policy position that the president should retain maximum discretion in the use of the SPR. The vagueness also accommodates conflicting views of the appropriate way to distribute SPR oil during drawdowns.

Conflict over the distribution question stems from divergent perceptions of how petroleum markets are likely to function during supply disruptions. On the one hand are those who believe that market forces, if left to operate, will minimize the social costs of disruptions by allocating scarce supplies to their highest valued uses. They view the regional storage of petroleum products as unnecessary and inefficient and favor selling SPR oil to the highest bidders during drawdowns. On the other hand are those who reject the notion that the market will allocate supplies efficiently and equitably during disruptions. They generally favor regional storage and believe SPR oil should be allocated to disadvantaged oil users at regulated prices.

This chapter reviews three issues related to the use of the SPR. The first is the question of the desirability of having a regional petroleum reserve (RPR) that would provide refined petroleum products directly to regional markets. The second is the method by which crude oil withdrawn from the SPR is to be distributed. The third, and conceptually the most interesting, is the specification of a strategy for determining when and at what rates the SPR should be drawn down.

The RPR Controversy

Sections 157 (a) and (b) of EPCA require the regional storage of refined products for which imports constitute 20 percent or more of regional consumption. Under this criterion the RPR would consist of residual fuel oil stored in New England and on the rest of the East Coast. However, Section 157 (c) of EPCA permits crude oil stored in a nearby region to be substituted for regional storage if it will result in comparable protection for the region. Although there has been internal disagreement and several apparent reversals of policy, the Ford and Carter administrations did not implement regional product storage despite strong pressure to do so from the New England congressional delegation.

As previously noted, the RPR question was one of the most difficult to resolve during preparation of the SPR Plan. Some analysts at FEA headquarters supported the position of the New England regional administrator, who stated that a reserve of residual fuel oil in New England was necessary to insure that the region would not suffer disproportionate economic harm while SPR crude oil stored in the Gulf Coast was being drawn down, refined, and transported.[1] Other FEA analysts, with OMB support, argued that lower-cost, centralized, crude oil storage would provide adequate protection for New England. After considerable debate, the SPR committee within FEA decided not to include implementation of the RPR in the SPR Plan.

RPR advocates challenged the FEA position during congressional hearings on the SPR Plan.[2] On the technical level, they argued that over 70 percent of the residual fuel oil consumed in New England was imported rather than the 58 percent claimed by FEA. They also pointed out that storage costs would be lower than assumed by FEA if No. 4 fuel oil, which does not require heating during drawdown, was stored in hard rock mines rather than if heavier residual fuel oil was stored in steel tanks.

At a more fundamental level, RPR advocates questioned the FEA conception of how the residual fuel oil market would function during disruptions. FEA assumed that in the estimated seven weeks it would take to convert SPR crude to fuel oil for New England, regional demand would be met by fuel oil in transit from Caribbean refineries and from other U.S. regions relying less heavily on imports. A summary of the opposing view is found in the statement of the chairman of the Energy Committee of the New England Council before the Senate Committee on Interior and Insular Affairs:

> The FEA plan tells us to prepare to depend on continued imports which are in transit, and on diversions from other areas of the United States, while Caribbean refineries continue to ship to us.
>
> During the last embargo, cargoes which were in transit at the time of the embargo were not delivered on time, at the usual port. FEA has documented that tankers were slowed or allowed to circle off the coast as the owners waited to determine how far the price would rise. We have no reason to believe that the same thing will not occur during the seven weeks it will take for the crude oil from the SPR to move into operations. In that period of time, New England could run out of residual fuel oil or be subjected to severe price gouging, unless it were willing to pay the high price which will certainly be demanded....
>
> Similarly we cannot simply depend on Caribbean refineries: they will react as would any other seller in the market, holding deliveries until the price is highest....
>
> Domestic refiners cannot be counted on as a source of fuel oil since they produce only half of the U.S. demand—and little of that product reaches the East Coast market.[3]

Although the Carter administration had indicated its intention to decontrol petroleum product prices, when the SPR Plan was being debated petroleum products were still subject to price controls as they had been during the Arab embargo, when traders reportedly delayed their deliveries to New England.[4] Conflicting views of petroleum product regulation lie at the heart of the RPR controversy.

The FEA position on the RPR implicitly assumes either the efficient operation of an unregulated market or the effective allocation of products by government in a regulated market. If it is assumed that suppliers of petroleum products attempt to maximize profits, the unregulated market would result in regional market-clearing prices that differed at most by transportation costs. To see this, imagine the price in region A was $50 per barrel and the price in region B was only $40 per barrel. A trader selling in region B could increase profits by shifting barrels to region A as long as transportation costs were less than $10 per barrel. In doing so, the trader would contribute to a slightly higher price in region B and a slightly lower price in region A. Only when the prices just differed by the difference in the cost of shipping to region A rather than region B would it not be possible to further increase profits. Acceptance of this view of the market leads to the conclusion that, despite the disproportionate reliance of New England on residual fuel oil imports in normal market periods, it would suffer roughly the same percentage loss of consumption as other U.S. regions having similar elasticities of demand. In fact, if New England demand is less elastic than other regions (as would be the case if it were more difficult for New England to switch to substitute fuels), it would suffer a smaller percentage loss in consumption.

If product prices are not allowed to rise to reflect regional scarcity, traders will have no incentive to shift supplies from their predisruption patterns of distribution. Government allocation of available supplies would be necessary to compensate for the impact of price constraints. In theory, the government could order movements of domestic supplies to equalize, in some sense, regional burdens. As long as domestic prices were kept below world prices, however, imports from nondisrupted sources would be diverted to markets outside of the United States.

The RPR advocates seemed to be drawing attention to the allocational problems inherent in price regulation while at the same time implying that price regulation is desirable. The delayed deliveries during the Arab embargo that they noted would have most likely been less pronounced if regulations had not restricted the rate of price increases. Advocates also wanted to know what assurances FEA had that Caribbean refiners provided SPR oil would not attempt to sell products back to the United States at world prices rather than at controlled prices.[5] Despite their recognition of these problems, they implicitly favored the continuation of price controls when they equated paying the world market price with being price gouged. Their arguments suggest that they viewed the RPR as a necessary complement to price regulation.

Senator Kennedy led the New England congressional delegation in an attempt to reverse the FEA decision. He introduced a resolution of disapproval of the SPR Plan but did not press it after receiving assurances from the FEA administrator that the new administration would continue to study the regional storage question and submit amendments to the SPR Plan in the future if warranted.

Within the Carter administration the RPR question became part of a larger battle over the SPR. James R. Schlesinger, the secretary of the newly created Department of Energy, favored the storage of at least ten million barrels of fuel oil in New England as part of the proposed expansion of the SPR to one billion barrels. In the fall of 1977 he included funding for implementation of the RPR in his FY 1979 budget request. The OMB cut these funds along with funds requested for expanding crude oil storage beyond 500 million barrels. Schlesinger appealed these cuts to President Carter. Because the analytical arguments for the RPR were not particularly strong, Schlesinger emphasized the political benefits of implementing at least some regional storage. The president accepted the OMB argument that the RPR was unnecessary and indicated that he did not wish to spend money on the RPR solely for political reasons.

The president's decision was reflected in Amendment No. 2 to the SPR Plan, which DOE submitted to Congress in April 1978.[6] The amendment called for the inclusion of regional storage in the expansion of the SPR to one billion barrels only if it would be no more expensive than the primary storage of crude oil. Senator Kennedy immediately recognized this provision as a rejection of the RPR. He filed a resolution of disapproval sponsored by all the New England senators. The Senate Energy Committee voted unanimously to back the resolution, and it was accepted on the Senate floor. The victory, however, was of little consequence. Although DOE was forced to resubmit the amendment with the offending language removed, the administration's position remained unchanged; in 1978 OMB cut DOE's request for funding of 10 million barrels of regional storage from the FY 1980 budget.

The administration also felt pressure from the House. During oversight hearings before the House Subcommittee on Energy and Power, administration witnesses were invariably interrogated about the RPR by Representatives Toby Moffett of Connecticut and Edward Markey of Massachusetts. Even though the DOE budget requests in 1977 and 1978 included funds for RPR implementation, DOE witnesses had to take the heat. As a former SPR office director later remarked, "Why a Democrat President fought a Democrat Congress on such an emotional issue, I could never understand.[7] It may be that the White House already viewed Senator Kennedy as a rival.

In April 1979 the administration reversed its position on the RPR in an attempt to gain support for its oil decontrol and windfall profits tax proposals. The White House staff recommended linking implementation of the RPR to passage of the Windfall Profits Tax. The administration announced that 24.3 million barrels of regional and noncontiguous storage would be developed if the Windfall Profits Tax was enacted and funds were available in the associated Energy Security Trust Fund. DOE established a task force to prepare an implementation plan, and the administration requested funding in the budget it submitted to Congress in January 1980.[8]

By March, however, the administration submitted a new budget without RPR funding. On the same day the Windfall Profits Tax was enacted, OMB sent instructions to DOE regarding the RPR: "As part of the President's proposal to balance the budget and reduce inflation, it was decided to defer the administration's program for regional storage in fiscal year 1980 to fiscal year 1981. Consequently, DOE should not submit any SPR plan amendment regarding regional storage to the Congress at this time."[9] In light of the history of the issue, Congress might have responded vocally and angrily to the reversal in policy. At the time, however, Senator Kennedy was on the campaign trail, and Representative Dingell's Energy and Power Subcommittee was concentrating its efforts on forcing the administration to resume filling the available crude oil storage capacity. Whether due to diverted attention or exhaustion, the congressional response was muted.

The final blow to the RPR came from within DOE itself. During the internal review of the department's FY 1982 budget, the Policy and Evaluation Office recommended elimination of funding for the RPR. Pointing to analyses of the Arab oil embargo and the Iranian supply crisis of 1979, they argued that it was unlikely that any one region would suffer a disproportionate loss of supply during a disruption and that hence the higher cost regional storage was unnecessary. As to the argument that U.S. refiners would be unable to meet residual fuel oil needs if all product imports were lost, they noted that the production of residual fuel oil could be dramatically increased if refiners simply reduced the level of downstream processing generally employed to convert resid (residual oil) to lighter products. Because of the nominal commitment of the Carter administration to the RPR, DOE left planning funds in the FY 1982 budget request. Even these funds were cut by the Reagan administration.

Distribution of SPR Oil

The original SPR Plan did little more than identify the key issues that would have to be addressed in developing a detailed distribution scheme. It noted that pricing and allocation procedures would have to be formulated to ensure an equitable distribution consistent with the provisions of the Emergency Petroleum Allocation Act (EPAA) and EPCA, the two legislative sources of standby price control and allocation authorities. Many congressmen, particularly those from the Northeast, desired more specific assurances that their regions would receive fair shares of the benefits of SPR drawdowns. Despite repeated requests from congressmen for completion of a detailed distribution plan during oversight hearings, the administration made little progress for almost two years. The SPR Office was faced with more pressing implementation problems, and the DOE leadership placed a relatively low priority on contingency planning.[10]

By the end of 1978, however, two groups within DOE were working on the distribution issue. One was the newly created Emergency Response Planning Office under the assistant secretary for Policy and Evaluation (PE). Its staff, which consisted largely of the former strategic planners for the SPR Office, favored selling SPR oil at the world price and allowing market forces to determine its distribution.

The other group was the Regulations and Emergency Planning Office of the Economic Regulatory Administration (ERA). Its orientation was more toward integrating SPR oil into the existing array of standby allocation authorities.

Beginning in January 1979, the two groups began working jointly on preparing an amendment to the SPR Plan. The first decision made was to establish basic sales agreements (BSA) with refiners who wished to be eligible to purchase SPR oil. The BSAs, issued beginning in May, set out basic contract terms intended to facilitate logistical planning for drawdowns. Within six months sixty-four refiners had entered into BSAs with the Department of Energy; participation has since risen to about 40 percent of U.S. refiners.

It was much more difficult to reach agreement on which BSA holders would be eligible to buy SPR oil and at what price. In the end these questions were not really resolved. The amendment transmitted to Congress on October 31, 1979, makes available a full range of options encompassing the market approach preferred by PE and the regulatory approach preferred by ERA.[11]

The first stage of the selection process set out in the amendment is to determine the eligible universe of recipients. Eligibility is to be limited either to BSA holders or to some subset of BSA holders specified by DOE for the purposes of promoting equitable access to available crude oil supplies, ensuring regional access to refined products, or facilitating the logistics of drawdown. In other words, DOE retained complete discretion over the extent to which SPR oil would be incorporated into existing or future crude oil and product allocation programs.

The second stage involves allocating oil among the eligible recipients by one of three methods: competitive sales, administrative determination according to announced criteria, or distribution of purchase rights on a pro rata basis according to historical use. Competitive sales, the preferred PE option, simply involves auctioning off SPR oil to the highest bidders. Under administrative determination, eligible refiners submit purchase orders at the DOE specified price. DOE then selects which purchase orders to honor according to the announced criteria. ERA favored allocation by administrative determination and had already proposed regulations giving DOE authority to set the criteria.[12] The third method, the pro rata distribution of purchase rights, was somewhat of a compromise between the PE and ERA positions. It allows DOE to administratively determine the price but restricts DOE authority to allocate to specific firms.

Even the administrative determination of price under the latter two methods is left virtually unconstrained. The amendment indicates that price will be set at approximately the average landed import price for similar quality crude oils, "unless it is determined that a lower or higher price is in the national interest."[13] Thus the amendment keeps the full range of regulatory options for distribution of SPR oil even in the absence of general regulatory authority after expiration of EPAA.

Several organizational factors contributed to the ERA victory. While the distribution plan was being prepared, Secretary Schlesinger's influence in the administration was declining, leading ultimately to his resignation. Whereas this change in leadership had relatively little impact on the semi-independent ERA, it marked a low point in PE's influence. At the time the Emergency Response Planning Office in PE was devoting most of its resources to analysis related to the crisis caused by the Iranian production cutbacks and literally relegated most of its work on the distribution plan to weekends. Finally, the political environment outside DOE favored the regulatory approach to contingency planning. Key Democratic congressmen and the White House domestic policy staff generally supported the use of price controls and allocation authorities during oil supply disruptions.

The SPR distribution issue was reexamined in the spring of 1981 as part of a general review of contingency planning policies. Although the Reagan victory knocked ERA out of the picture, a number of former ERA staffers were able to secure positions with the newly created Office of Environmental Protection, Safety, and Emergency Preparedness, which gained jurisdiction over the SPR Office. These "boat people," a name given to them by opponents in other parts of DOE, attempted to keep the regulatory approach as an option.[14] They argued that allocation of SPR oil might be necessary to guarantee small and independent refiners access to crude oil during major disruptions. Major refiners hold long-term contracts with producers that would enable them to secure some supplies at lower than spot market prices during disruptions. The majors might use the profits from these supplies to outbid small and independent refiners in the spot market and SPR auctions. They further argued that the Department of Defense and other regular customers of the small and independent refiners might not be able to find alternative sources of supply.

Exclusive use of the auction approach was urged by the Office of Policy, Planning and Analysis (PPA), which replaced PE in the DOE reorganization. The PPA analysts thought it unlikely that the majors would be willing to bid more for SPR oil than the value of the products it will yield. Doing so would lose profits and increase the risk of political intervention in the oil market. Refiners who are inefficient or unable to alter their product slates to meet changes in demand might very well not be able to purchase oil in the spot market or at SPR auctions at profitable prices, but this is appropriate in terms of maximizing the aggregate economic value from the use of available oil supplies. As long as the government refrained from interfering in product markets, the pursuit of profits by refiners and traders would result in product availability across regions of the country and sectors of the economy. If paying market-clearing prices placed undue economic burdens on specific groups of consumers, the government could provide relief in the form of grants or tax reductions.[15]

PPA analysts made two other arguments against the regulatory option. First, potential beneficiaries of regulation must understand that it will not be implemented so that they will have an incentive to take actions to protect themselves. For example, small and independent refiners might form purchasing groups to secure oil supplies under long-term contracts so that they will not feel the full impact of increases in spot market prices.[16] However, the incentive to do so will be much weaker if refiners believe they will be given preferential access to SPR oil at regulated prices during disruptions.

Second, allocation of SPR oil would increase the likelihood of a return to government allocation of all crude oil supplies. Once the government begins giving special treatment to the refiners serving certain groups of consumers, political pressure will mount to give equal treatment to other groups that are deserving or politically powerful. Because these demands would undoubtedly exceed SPR drawdown rates, the political pressure could easily translate into reimposition of the Buy/Sell Program or other regulatory authorities. PPA believed that a comprehensive market approach would be much easier to defend against claims for special treatment than an approach that already opens the door to regulation.

Although the position of the advocates of the regulatory approach was bolstered somewhat by a report of the National Petroleum Council advocating government allocation of oil during severe disruptions, energy analysts in OMB, CEA, and the White House generally supported the market approach.[17] The compromise finally hammered out between the Office for Environmental Protection, Safety, and Emergency Preparedness and PPA set a presumption that SPR oil would be distributed by auction but kept open the possibility of noncompetitive sales to refiners supplying essential products to meet defense needs, to refiners serving geographical regions severely affected by the disruption, or to refiners meeting U.S. obligations under the International Energy Program administered by the International Energy Agency. The compromise was accepted by the cabinet council on natural resources and environment, the energy policy clearing house for the Reagan administration.

Drawdown Strategy

EPCA permits the withdrawal and distribution of oil in the SPR upon a finding by the president that a drawdown is required either to counter a "severe energy supply interruption" or to meet U.S. obligations under the International Energy Program. EPCA defines "severe energy supply interruption" as:

> A national energy supply shortage which the President determines—
> A. is, or is likely to be, of significant scope and duration, and of an emergency nature;
> B. may cause a major adverse impact on national safety or the national economy; and
> C. results, or is likely to result, from an interruption in the supply of imported petroleum products, or from sabotage or an act of God.[18]

This definition is sufficiently vague to allow the president to declare a "severe energy supply interruption" at the onset of even relatively small disruptions of the world oil market. Further, EPCA does not specify conditions under which the president must draw down the SPR. In other words, EPCA gives the president almost total discretion over the initiation of drawdowns.

The president's discretion may be limited through provisions of the SPR Plan. However, neither the original SPR Plan nor the 1979 distribution amendment placed any limitations on, or requirements for, the initiation and rates of drawdowns. In particular, there is no specification of circumstances that would automatically trigger an SPR drawdown.

The establishment of trigger mechanisms has been occasionally advanced, including most recently by analysts in President Reagan's Council of Economic Advisors. The trigger mechanisms would reduce presidential discretion linking the initiation and perhaps even the rates of drawdowns to the occurrence of specified conditions in the world oil market. Sudden reductions in import levels or supply to the world market and steep increases in spot market prices have been recommended as triggers. Tying drawdown rates to changes in private sector stock levels or drawdowns by other members of the International Energy Agency has also been suggested.

A variety of arguments have been advanced in support of the establishment of trigger mechanisms. The one most frequently advanced reflects skepticism about the prospects for economically efficient drawdown decisions in the political environments that accompany disruptions. The greatest concern is that the president and his advisors will be hesitant to initiate drawdowns. Whereas the economic benefits from use of the SPR are probably greatest at the onset of disruptions, when consumers have not yet had a chance to adjust to higher prices, fear that the disruptions will be more severe or of longer duration than expected may lead decision makers to delay the initiation of drawdowns until more information is available. Appearing to have used the reserve too quickly is likely to be more politically costly than appearing to have been overly cautious in its use. More information and analysis is necessary to increase the chances of avoiding the former error; but this can result in delays that make the latter more likely. A number of analysts have advocated putting trigger mechanisms in place prior to disruptions as a way of eliminating the political bias toward underutilization.

Other analysts have argued that trigger mechanisms are desirable because they reduce the uncertainty about how the SPR will be used.[19] The private sector will be less likely to stockpile or to institute other types of self-protection if there is a likelihood that the SPR will be used to counter the price rises caused by minor disruptions. If a policy is set limiting SPR drawdowns to major disruptions, formalizing it through the establishment of trigger mechanisms would give the private sector greater confidence that it will be followed. Similarly, the establishment of trigger mechanisms might prove useful in reassuring allies that the United States is willing to actually use the SPR.[20]

The major argument against the establishment of trigger mechanisms is that no simple set of rules can adequately take account of all the factors relevant to drawdown decisions. An unambiguous assessment of current market conditions is often difficult to make from available data. More importantly, none of the trigger mechanisms that have been proposed adequately take account of expectations about future market conditions. The myopia is particularly serious for triggers that are based on production or import levels. Estimates of recent changes in these levels provide virtually no information about future changes.

The myopia is less serious for triggers based on observations of market prices. Spot prices and newly negotiated long-term contract prices do incorporate private sector expectations about future market conditions. However, spot prices are dominated by current supply and demand conditions, and official long-term contract prices are slow to change and are subject to manipulation. The best information about private sector expectations would be provided by the prices observed in full-fledged crude oil futures markets. Unfortunately, such markets do not now exist and, if they did, might not continue to operate efficiently in the face of severe disruptions or manipulations by major producers.

Other objective triggers suffer from the same myopia. Expectations about future market conditions must be formed subjectively from a variety of market data and intelligence sources. Trigger advocates, while generally recognizing the conceptual importance of expectations to the determination of optional drawdown decisions, are skeptical about the capacity of SPR decision makers to effectively use such subjective information. Those who oppose the institution of trigger mechanisms generally place greater emphasis on the importance of expectations and tend to be more optimistic about the role of subjective information in the decision-making process.

While rejecting the notion of triggers, a number of analysts have proposed strategies that would guide the exercise of discretion by decision makers. The first strategies proposed were based largely on intuition. For example, the benefit-cost analysis presented in the SPR Plan assumed an exponential drawdown pattern would be employed as a hedge against disruptions lasting longer than expected. The maximum drawdown each day would be limited to some fixed percentage of the total amount left in the reserve on the previous day. The exponential drawdown rule would result in a gradual decline in the amount of oil released from the reserve as the disruption dragged on.[21]

The first formal modeling of the drawdown problem was completed by analysts at the Institute for Defense Analyses.[22] In work sponsored by the FEA, they attempted to find the optimal drawdown pattern under a variety of assumptions about the relationship between oil import reductions and gross national product. Although their analysis assumed that the duration of the disruption is known with certainty, it allowed for the probability of the occurrence of additional disruptions in later periods. It thus constituted a first step toward formally taking account of expectations about future market conditions.

A more sophisticated treatment of expectations is embedded in the dynamic programming model of the SPR developed by the Policy and Evaluation Office (PE) of DOE in 1979. As previously discussed, the PE model was injected into the debate over the appropriate size of the SPR. The initial motivation for its development, however, was the determination of optimal acquisition and drawdown policies for the SPR.

The model assumes that the world oil market will be in one of a number of identifiable states in the current and each of the future time periods.[23] For example, the model might employ five distinct market states: two nondisrupted states (tight and slack markets) and three disrupted states (minor, moderate, and major supply reductions). Expectations are built into the model through the assumption that there are fixed probabilities of moving from each market state to each of the other market states. The matrix of the transition probabilities summarizes expectations about future market conditions.

The specification of the transition probabilities introduces risk into the calculation of the net benefits of alternative drawdown and acquisition decisions. By assuming nonzero probabilities of staying in the same disrupted market states or moving to other disrupted states, the model reflects the reality that the course of disruptions cannot be known with certainty at their start. In calculating the present value of expected net benefits of alternative drawdown and acquisition decisions, the model takes account of the range of possible market transitions that could occur in the future. The particular decision that maximizes the present value of expected net benefits will be a function of the amount of oil already in the SPR, the current state of the market, and the expectations about future market conditions as reflected in the assumed transition probabilities.

Analysis based on the model suggests that the SPR will provide the greatest net benefits if acquisition and drawdown decisions are constantly reassessed to reflect the amount of oil in reserve, current market conditions, and expectations about future market conditions.[24] Filling at a constant, moderate, rate until planned capacity is full and initiating drawdowns only during major disruptions will not generally be optimal policies. For all but the most optimistic expectations about future market conditions, the model suggests that the SPR should be filled at the maximum feasible rate (currently about 500,000 barrels per day) during slack markets. Continued fill during tight markets and drawdowns during minor disruptions may or may not be optimal, depending upon the current amount of oil in the SPR and expectations about future market conditions.

In practice could a model similar to the one developed by PE be used to guide the execution of a more nearly optimal acquisition and drawdown strategy? Our previous discussion of the 1980 fill controversy (chapter 4) does not bode well for the implementation of a flexible acquisition strategy under current institutional arrangements, where short-run budgetary and economic costs tend to overshadow future expected benefits in the political calculus. The direct usefulness of such a model in guiding decisions concerning drawdowns is likely to be severely limited by the difficulty of translating the expectations of decision makers into probabilities. Our discussion of the role of analysis in the bureaucratic battles over the development of SPR storage capacity (chapter 6) suggests that agreement on reasonable probability estimates will be difficult to reach if the analysts involved happen to hold conflicting preconceptions of appropriate policies.

Aside from contributing to a general understanding of the factors that should be considered in making drawdown decisions, the model might be a useful tool in countering the possible bias toward underutilization. Rather than finding the optimal decision for a given set of expectations, the model might be used to find the set of expectations the decision maker would have to hold to make not drawing down the reserve an appropriate decision.[25] In this way, decision makers would receive an indication of how likely they must believe it to be that the disruption will last longer or be more severe than expected.

Summary

Progress in the resolution of distribution and drawdown issues has been slow. It now appears almost certain that there will be no regional storage of petroleum products; the endurance of RPR proponents finally seems to have been dissipated. Although the Reagan administration has established a strong presumption that SPR oil will be distributed through auction, formal rules have not been put in place as insulation against those interests that will undoubtedly seek preferential access during disruptions. There is still no meaningful drawdown policy. While trigger mechanisms are too blunt, total discretion will leave decision makers subject to intense conflicting pressure during disruptions. Unless policies are put into place prior to the next disruption, there is a danger that these pressures will lead to far from optimal decisions.

Notes

1. The regional administrator of FEA was chairman of the Energy Resource Development Task Force of the New England Federal Regional Council, which completed a report entitled "Energy Petroleum Storage in New England," in October of 1976. U.S. Congress, Senate, "Review of the Strategic Petroleum Reserve Plan," Hearing before the Committee on Interior and Insular Affairs, 95th Cong., 1st sess., February 4, 1977, pp. 131–230.

2. "Review of the Strategic Petroleum Reserve Plan"; and U.S. Congress, House, "Strategic Petroleum Reserves," Hearing before the Subcommittee on Energy and Power of the Committee on Interstate and Foreign Commerce, 95th Cong., 1st sess., February 16, 1977.

3. Prepared statement of Zeb D. Alford, chairman, Energy Committee, New England Council before the Senate Committee on Interior and Insular

Affairs, "Review of Strategic Petroleum Reserve Plan," February 4, 1977, pp. 117–20.

4. The Phase IV controls instituted by the Cost of Living Council were in effect during the Arab embargo. Refiners and importers could pass through to consumers most of their higher crude oil and imported product costs, but only on an average basis. Additional price constraints applied to gasoline, diesel oil, and No. 2 fuel oil; the latter two products are close substitutes for residual fuel oil. Although banking of credits for sales at below ceiling prices provided some flexibility to the system, the pricing rules employed did not permit instantaneous pass-through of cost increases. It is therefore not surprising that traders delayed their deliveries to allow ceiling prices to catch up and to avoid government allocation of their inventories.

5. This issue was raised in the prepared statement of John H. Lichtblau before the House Subcommittee on Energy and Power, "Strategic Petroleum Reserves," February 16, 1977, p. 201.

6. Strategic Petroleum Reserve Office, "Strategic Petroleum Reserve Plan Amendment No. 2 (Energy Action DOE No. 1): Expansion of the Strategic Petroleum Reserve," March 1978.

7. Interview with Jay R. Brill, February 19, 1981.

8. Department of Energy, Task Force on Regional and Noncontiguous Storage, "Program/Project Plan for Regional and Noncontiguous Petroleum Reserve Storage Project," Preliminary Report, May 30, 1980.

9. Memorandum from Hugh Loweth to Jack Hewitt entitled "SPR Budget Policy Guidance," March 27, 1980. U.S. Congress, House, "Filling the Strategic Petroleum Reserve: Oversight; and H.R. 7252: Use of Naval Petroleum Reserves," Hearing before the Subcommittee on Energy and Power of the Committee on Interstate and Foreign Commerce, 96th Cong., 2d sess., April 25, 1980, pp. 49-51.

10. As concluded by the U.S. Congress, Senate, Committee on Governmental Affairs, "Oversight of the Structure and Management of the Department of Energy," Staff Report, December 1980 (Washington, D.C.: Government Printing Office, 1980), pp. 260–61.

11. U.S. Department of Energy, Assistant Secretary for Resource Applications, "Strategic Petroleum Reserve Plan Amendment No. 3 (Energy Action DOE No. 5): Distribution Plan for the Strategic Petroleum Reserve," October 1979.

12. Economic Regulatory Administration, "Notice of Proposed Rulemaking: Distribution of Strategic Petroleum Reserve Crude Oil," *Federal Register,* August 20, 1979, pp. 48696–707.

13. "Strategic Petroleum Reserve Plan Amendment No. 3," pp. 7-8.

14. See Richard Corrigan, "The Next Energy Crisis—A Job for the Government or Free Market?" *National Journal,* June 20, 1981, pp. 1106-9.

15. These ideas were set out in detail in a contingency planning report prepared by the Office of Energy Security in PPA. See U.S. Department of Energy, Office of Energy Security, "An Emergency Preparedness Strategy for Oil Supply Disruptions," Draft, May 30, 1981.

16. Such purchasing groups have already been formed. See Burt Solomon, "Strength in Numbers: Independent Refiners Seek Own Access to Crude," *Energy Daily*, March 30, 1981, p. 3. For a discussion of independent refinery questions, see Edward J. Mitchell, "Protection for Petroleum Refiners," *Regulation*, July/August 1981, pp. 37-42.

17. National Petroleum Council, "Emergency Preparedness for Interruptions of Petroleum Imports into the United States," April 1981. The recommendation that the government allocate crude oil during only severe disruptions was far from unanimous. Some independent refiners favored allocation for smaller disruptions as well, while many members opposed government allocations altogether. Some participants claimed that the recommendation was the result of efforts by several of the majors who are heavy importers of oil from Saudi Arabia. Because the most likely scenarios for severe disruptions involve loss of Saudi production, these companies could expect to be recipients of allocations under the policy.

18. Energy Policy and Conservation Act, Public Law 94-163, Title I, Section 3 (8).

19. One advantage of uncertainty is that it might increase the difficulty adversaries face in planning an embargo or other disruption of the world market. This point was raised in the SPR Plan as an argument against the establishment of trigger mechanisms. "SPR Plan," December 15, 1976, pp. 149-50.

20. Some analysts fear that benefits of U.S. drawdowns will be reduced through attempts by other IEA members to increase their stocks at the onset of disruptions. Linking the rate of U.S. drawdowns to the drawdown rates of other IEA members has been suggested as a way to force a coordinated response to disruptions. Unfortunately, it would involve the risk of there being no drawdowns in situations where even U.S. drawdowns alone would provide large benefits to the U.S. economy.

21. For example, assume the SPR contains 500 million barrels, has a technical limit on drawdown of 3 million barrels per day, and is to be used to counter a disruption that is expected to last 150 days. A linear drawdown rule might call for a drawdown of 3 million barrels per day until either the disruption ends or the reserve is exhausted. If the disruption lasts longer than 166 days, the reserve will be exhausted, and the world market will suffer the shock of the loss of 3 million barrels per day of supply. An exponential drawdown rule set at 1 percent would result in a 3 million barrels per day drawdown for the first 66 days followed by a gradual decline in the draw-

down rate. At 166 days, the drawdown rate would still be in excess of one million barrels per day.

22. Results are reported in Robert E. Kuenne, Jerry W. Blankenship and Paul F. McCoy, "Optimal Drawdown Patterns for Strategic Petroleum Reserves," *Energy Economics*, January 1979, pp. 3-13.

23. For instance, in the stylized example presented in chapter 5, there were only two market states: disrupted and nondisrupted.

24. Glen Sweetnam et al., "An Analysis of Acquisition and Drawdown Strategies for the Strategic Petroleum Reserve," Draft, U.S. Department of Energy, Assistant Secretary for Policy and Evaluation, Office of Oil, December 17, 1979.

25. This is the approach anticipated by the Office of Energy Security in PPA (formerly PE). Although analysts in the SPR Office have begun to use the PE model for investigating capacity development questions, they have not yet started using it as a tool for guiding acquisition and drawdown decisions.

PART III

CONCLUSION

9

Perspectives on SPR History

Because of the importance of the strategic petroleum reserve program to national security, simply recording its history is intrinsically worthwhile. Indeed, some may have read the preceding chapters solely out of curiosity about the program and how it came to be what it is. Hopefully, these readers have not been disappointed. However, this book was written primarily for students of the policy sciences including, but not limited to, those specifically interested in energy policy. In particular, the history was analyzed from four perspectives: First, what approaches for improving the SPR program are suggested by its history? Second, how does the SPR program relate to the development of U.S. energy policy in general? Third, what can be learned from the history of the SPR program about the implementation of new programs by the federal government? Fourth, what does the history of SPR decision making suggest about the use and abuse of formal analysis in bureaucratic settings? The purpose of this concluding chapter is to highlight briefly these perspectives.

Improving Stockpiling Policies

The SPR program is no longer in crisis. After initial years of schedule slippage and cost overruns, it is now meeting less ambitious but more realistic goals and, as a result, restoring its credibility. Although about four years behind its initial schedule, storage capacity is now being added at a slow but steady pace. After a pause of more than a year, available capacity is being filled at a relatively fast rate. While recruiting and keeping personnel with critical managerial and technical skills remain problems under the civil service system, capabilities are improving with the accumulation of experience and the strengthening of the managerial infrastructure. If morale is not as high as it was in the early days under the FEA, it is at least substantially higher than at the low point in 1979.

Despite these improvements in the SPR program itself, its effectiveness in reducing U.S. vulnerability to disruptions of the world oil market will depend greatly on the way a number of policy issues are resolved over the next few years. Four issues are particularly important: the organizational position of the SPR Office, the development of Phase IV storage capacity, the financing of the program, and the formulation of drawdown policies.

The dismemberment of DOE proposed by the Reagan administration provides an opportunity for reassessing the organizational structure and location of the SPR Office. Simply moving the SPR to another department, however, offers little or no advantage over the status quo. For example, as part of the Department of Defense its mission might shift from being a reserve for use during any major disruption of the world oil market to being a reserve for use only during major wars. The SPR Office would be an anomaly at the Department of Commerce. At either the Department of Interior or the Federal Emergency Management Agency, which oversees the critical materials program, the SPR Office would be in an environment not radically different from the one it is in at DOE.

In the past dissatisfaction with the handling of the SPR program by DOE has prompted calls for the establishment of the SPR Office as an independent agency or public corporation. As such it might be less subject to short-run budgetary considerations than if it were part of a larger department responsible for numerous programs with more vocal constituencies. However, the greatest advantage would result from moving it out from under the civil service system that makes it difficult to maintain a highly qualified staff. In light of the problems actually encountered, it is not unreasonable to believe that implementation of the SPR program would have been much more successful if the SPR Office had been originally constituted as an independent agency or public corporation.

The development of storage capacity by the SPR program remains problematical. After several years of delay due to bureaucratic politics, development of Phase III (the third 250 million barrels) storage capacity has finally begun but will not be completed until the end of the decade at the earliest. Phase IV, which would increase storage capacity to one billion barrels, is now under consideration. If a budgetary commitment is made in 1982, Phase IV could be completed by the end of 1990 through the creation of additional solution-mined salt dome caverns or somewhat earlier through a combination of salt dome caverns and more expensive steel-tank storage.

The weight of analysis currently favors initiation of Phase IV. Although future analysis might argue against completion of Phase IV, the long lead times for construction require work to begin immediately if the billion barrel reserve is to be an option for this decade. Meanwhile, the search should continue for ways of making storage capacity available on a faster schedule. In particular, purchase of privately developed storage capacity might substitute for direct government development of a portion of Phase IV. Cancellation of the turnkey initiative in 1979 was probably a mistake; similar opportunities should not be overlooked in the future.

Alternatives to direct budgetary financing of the SPR program were seriously considered for the first time amidst the budget cutting of 1981. Although the decision to move SPR financing off budget in an accounting sense is of little substantive importance, two rejected alternatives deserve reconsideration. One is the financing of the SPR program through a tariff on imported crude oil and petroleum products; the other is equity financing through the sale of oil-backed certificates with values tied to the world price of oil.

Many economists believe that the social costs to the United States of the marginal barrel of imported oil exceeds its world market price by an "import premium" of several dollars or more. If this belief is correct, a tariff equal to the import premium would increase economic efficiency by making consumers face the full social cost of oil consumption. Setting the tariff at about two dollars per barrel would cover SPR program costs at current levels of activity. The tariff could also be viewed as a user charge: it is paid by consumers of petroleum who in turn would be major beneficiaries of SPR drawdowns. Thus financing the SPR through a tariff on crude oil and petroleum product imports appears attractive in terms of both economic efficiency and fairness.

Equity financing, although more difficult to implement, likewise offers benefits beyond simply moving the SPR program off budget. It would provide an investment opportunity for the private sector to diversify against the risk of disruptions. It would help stabilize the distribution of national income during disruptions, reducing macroeconomic losses. Finally, it would create a political constituency against price controls and for drawing down the SPR.

Little progress has been made toward formulation of effective drawdown policies. There is likely to be a political bias toward underutilization of the SPR. Unfortunately, it appears that there is no satisfactory trigger mechanism for countering the bias toward underutilization. Another approach that deserves consideration is the decentralization of decision making about drawdowns for a fraction of the SPR. For example, the government might set aside Phase IV storage capacity for use by the private sector on a fee (or subsidy) basis. Aside from minimum holding periods to discourage panic buying at the onset of disruptions, the private owners of the oil would be permitted to buy or sell their oil as they see fit. The volume of oil in the capacity reserved for the private sector would change in response to changes in private sector expectations.

Relation to U.S. Energy Policy

Since 1973 U.S. energy policy has been politically salient and controversial. In both direct and subtle ways, energy policy touches our lives as consumers, workers, and investors. It has important links to our foreign policy and national defense. It often involves tradeoffs between economic efficiency and the distribution of wealth that bring organized interests into conflict. It must be formulated by Congress and the president amidst conflicting information, analyses, and ideologies. Its impact will be influenced by the way it is interpreted and implemented by bureaucrats with personnel preferences and organizational interests. These factors make energy policy an interesting and important topic for study.

A variety of approaches for exploring the development of U.S. energy policy have been employed. For example, David Howard

Davis in *Energy Politics* provides an historical overview of the politics associated with major energy sources.[1] The Brookings Institution volume edited by Craufurd D. Goodwin traces the evolution of U.S. energy policy through successive postwar administrations, providing a rich description of executive branch decision making.[2] Thomas H. Tietenberg concentrates on policy formation immediately after the 1973 Arab embargo, and Walter A. Rosenbaum considers the general factors influencing energy policy formation.[3] *The Politics of Mistrust* by Aaron Wildavsky and Ellen Tenenbaum considers the role estimates of oil and natural gas resources have played in the history of U.S. energy policy and politics; it is closest in approach to our investigation of the planning and analysis behind the SPR program.[4]

One thesis of Wildavsky and Tenenbaum is that much of the history of controversy over U.S. energy policy can be understood in terms of conflicts between two general perspectives on the nature of the value of energy resources to society. They describe one perspective as a "quantity theory" view of the world: energy resources are finite and therefore have intrinsic values that do not necessarily correspond to their market prices. Further, the government should take actions to ensure that future supplies are adequate to meet future needs. In contrast, the "price theory" view sees the prices resulting from the interaction of self-interested producers and consumers as the best indication of the value of energy resources to society. If the government is going to intervene, it should do so to make energy markets more efficient. Whereas quantity theorists see energy resources as unique assets that must be guarded for future generations, price theorists view them as interchangeable in the long run with other economic inputs.

Over the last decade the quantity theory has dominated U.S. energy policy. Reduction of dependence on imported oil through nonprice mechanisms became the dominant goal. Coercive regulations were brought to bear in an attempt to make consumers conserve, utilities switch to coal, and automobile manufacturers produce more efficient vehicles. The federal government moved toward an active role in promoting nuclear power, solar energy, and synthetic fuels. Recognizing the currency of the quantity theory perspective makes it only somewhat less paradoxical that during the same period well-head price controls on crude oil and natural gas were contributing to greater reliance on foreign oil.

The distinction between vulnerability and dependence arises out of the price theory perspective. In and of itself, dependence on foreign oil is no worse than dependence on any other imported commodity. Dependence is a serious problem primarily because the world oil market is vulnerable to disruption. Social benefits will result from U.S. policies that reduce the cost of disruptions in the world oil market. Reducing levels of oil imports will also produce social benefits by reducing transfers of wealth to foreign oil producers during price shocks. Preparing contingency measures, including stockpile drawdowns, produces social benefits by limiting the magnitude of price shocks or the harm they cause to the economy. The important question is which approach or combination of approaches yields the greatest net social benefits.

Beginning with the Project Independence Report, analyses accumulated pointing to stockpiling as the most cost-effective approach for reducing the vulnerability of the United States and its allies to disruptions of the world oil market. The Ford administration and Congress, recognizing the desirability of stockpiling, established the SPR program with an ambitious timetable of goals. The SPR program enjoyed a high priority within the Federal Energy Administration, which translated into direct access to the administrator and therefore fast action on personnel and other resource requests. Mistakes in strategic planning were made. Nevertheless, the organizational environment facilitated the correction and accommodation of errors. Had the organizational environment continued to be favorable, implementation would undoubtedly have been more successful.

By accelerating its already ambitious schedule and expanding its planned size, the Carter administration seemed to be giving the SPR program an even higher priority. However, there was no accompanying increase in the organizational resources available for achieving the new goals. In fact, the creation of the Department of Energy eliminated many of the organizational advantages enjoyed under the Federal Energy Administration. Access to high-level administration decision makers was reduced, the orderly recruitment of an appropriately skilled staff was hindered, and emphasis was shifted away from contingency planning toward import reduction.

Although the quantity theory perspective leads to import reduction programs that compete with the SPR for scarce resources, it embraces stockpiling as a short- and medium-range necessity. Rather than moderating the magnitude of price shocks to the world economy, SPR drawdowns are viewed as directly replacing quantities of lost imports. In early studies this view led analysts to ignore the likely leakage of some fraction of drawdowns to the world market through the displacement of imports. Of greater relevance to the development of SPR policy, the quantity theory perspective led analysts to argue in favor of the establishment of regional reserves and the distribution of SPR drawdowns through direct allocation at administered prices.

With the notable exception of the promotion of nuclear power, the Reagan administration is realigning U.S. energy policy to conform more closely to the price theory perspective. The administration has rejected standby price control and allocation authorities, emphasizing instead the expeditious filling of the SPR. In addition to providing economic benefits, SPR drawdowns are seen as affirmative steps that can be taken to diffuse political pressure for nonmarket responses to disruptions. Only time will tell if the Reagan administration's commitment to the SPR will remain strong in the face of persistent budget deficits.

Implementing New Programs

Usually the responsibility for implementing new federal programs falls on organizations that have experience with similar programs already in existence. Occasionally, however, programs are created coincidently with the organizations that must implement them. In such circumstances, implementation cannot be separated from organizational development.[5]

The Energy Policy and Conservation Act gave FEA one year to complete the basic strategic planning for the SPR. FEA had to first recruit the staff who would gather and analyze information for planning decisions. The handful of personnel already on board had very limited knowledge of the technical issues that would have to be confronted. It proved difficult to quickly add personnel with specialized technical and mangerial skills. Heavy reliance had to be placed on contract research, much of which was completed by firms with only limited experience in dealing with similar problems.

Nevertheless, the strategic plan was completed on time. Although sound in most aspects, the plan had one serious flaw; it established an unwieldy system for managing design and construction contracts. The flaw would have been of less consequence, or perhaps even avoided in the first place, if the SPR Office included more persons with prior experience in managing design and construction (as opposed to research) contracts. The civil service system and external events limited the rate at which such persons could be added to the staff. The decision to begin execution of project contracts prior to completion of systems design work and the development of an adequate management control infrastructure meant that those already on the staff did not have an opportunity to acquire gradually supplementary expertise.

In light of the problems encountered during start-up of the SPR program, what are the generic problems likely to be encountered during implementation of other new programs involving unfamilar technologies and requiring expeditious completion?

Aspects of the program may come under jurisdiction of existing bureaus or regulatory agencies. When technologies are being employed in novel ways, these organizations are unlikely to have precedents that could facilitate rapid decision making. The result may be delay for procedural reasons. For example, the inability of the SPR Office to obtain brine discharge permits from the Environmental Protection Agency on a timely basis was a major source of delay. This problem might have been avoided if the enabling legislation had included provisions requiring relevant regulatory decisions to be made within a certain period or superseded them entirely by the congressional review of the plan. Under the actual legislation, the problem might have been mitigated if the DOE leadership had given greater support to the SPR Office in its dealings with the Environmental Protection Agency.

Recruiting and keeping appropriately skilled personnel is a problem facing all organizations, private or public. It is particularly serious for new programs developing under the civil service system. Government pay schedules may not be competitive with the private sector, particularly for persons with highly specialized skills. The SPR program continues to have difficulty attracting experienced personnel away from the high-paying oil industry. Because the civil service system is geared toward hiring permanent employees, it moves slowly. Once qualified people are found, it may take several months to actually hire them. Additionally, once they are hired, it is very difficult to replace them with others as the expertise needs of the organization change. This permanence, coupled with lower than anticipated personnel ceilings, meant that the SPR program had to begin design and construction without an adequate number of persons with technical and managerial expertise because scarce personnel slots were already occupied by persons previously needed for their contributions to strategic planning.

Private firms do not face the permanence problem. In fact, one way public organizations avoid permnanently collecting persons whose skills are only temporarily needed is to contract work out to private firms. A skilled staff is still required, however, to write effective contracts and monitor their execution. As the program matures and experience accumulates, the feasibility of contracting out larger shares of program activity will increase.

The personnel problems encountered by new programs would be reduced by civil service reforms that make it easier to hire personnel for limited periods at wages above scale. (Something skillful bureaucrats often accomplish through the creative use of contractors.) Perhaps the simplest approach would be to allow directors of new programs to use a fraction of their budgets to hire outside the civil service system during the first year or two of implementation. Although the potential for abuse would be high under such a system, it would allow program directors to function more like their private sector counterparts.

Another approach would be to have a central pool of analysts from which personnel could be borrowed until appropriately skilled permanent employees are recruited. This would allow program directors to reserve personnel slots for persons whose skills will be needed during later stages of implementation. However, it may be unrealistic to expect that a qualified pool could be maintained. Agencies would tend to hire away persons of above-average quality who are loaned to them, while sending those of below average quality back to the pool. Placing new programs under existing departments with large staffs keeps open the possibility of intradepartmental loans of personnel. Unfortunately, the new program usually finds itself in competition for resources, including the attention of decision makers, with established departmental interests, as did the SPR Office under the Department of Energy.

The Role of Formal Analysis in Bureaucratic Settings

The fundamental concerns of the emerging discipline of policy science are the actual, potential, and desirable roles for formal analysis in public decision-making processes. An extensive literature deals with these concerns from the general perspectives of the utilization of social science research and the development of the policy analysis profession.[6] Our study of the role of quantitative economic analysis in SPR decision making adds to a relatively smaller literature focusing explicitly on the use of formal analysis in organizational settings.[7]

Formal economic analysis was a factor of only minor importance in the initial decisions concerning the size of the SPR. Although within the then available range of estimates of the economically efficient size, the 500 million barrel goal set by the Ford administration represented a compromise between FEA and OMB positions. The decision by the Carter administration to increase the goal to one billion barrels was made almost exclusively on the basis of national security arguments. In an attempt to overturn or at least delay implementation of the latter decision, OMB maneuvered DOE into the position of having to justify an expanded SPR with economic analysis.

Once DOE accepted responsibility for showing the billion barrel reserve to be justified on economic grounds, OMB was in a position to delay implementation by casting doubt on the DOE analysis. When the mutually agreed upon methology produced results supporting the billion barrel goal, OMB attempted to discredit the analysis by challenging assumptions it had previously accepted. OMB forced repetition of the analysis in successive budget cycles. As long as the analysts could not reach a consensus on the technical issues, the DOE leadership was unwilling to appeal OMB's budget cuts to the president.

In his essay on OMB and the presidency, Hugh Heclo warns of the danger of "gratuitous conflict" if OMB does not show neutral competence in advising the president.[8] Forcing repetition of the SPR size studies appears to have been gratuitous. After the first joint size study, DOE analysts realized that their OMB counterparts would not be satisfied with an analytical product supporting expansion of the SPR beyond 500 million barrels. Participants from the Council of Economic Advisors might have provided neutral competence for resolving technical disputes, but they did not.

The SPR studies suggest that without procedures for reaching a binding consensus on analytical assumptions prior to the determination of results, studies jointly conducted by bureaucrats with radically different policy preferences are unlikely to be deciding factors in policy disputes. In such situations, those advocating action must seek resolution of the issue at a higher political level to escape the "paralysis of analysis." This approach was taken by the new DOE assistant secretary for Policy and Evaluation in 1980 after he became convinced that the analysis completed by his staff was sound. Although other factors may have been important as well, the threat of taking the issue to the president undoubtedly contributed to OMB's decision not to cut Phase III funding from the SPR budget as it had done in the three previous years.

Did analysis make a difference? Yes, but in indirect ways. If the studies DOE conducted jointly with OMB had shown expansion of the SPR to be unjustified on economic grounds, expansion would probably have been permanently ruled out. Without the cumulative weight of these studies and, more importantly, the analysis conducted independently within the Office of Policy and Evaluation, the assistant secretary probably would not have been willing to bear the costs of urging an appeal to the president. Though it is rarely decisive, analysis is one of many resources that can be used and abused in bureaucratic politics.

If the president and Congress recognized that expeditious development of a large petroleum stockpile is in the public interest, why was bureaucratic politics allowed to become such an important factor in the implementation of the SPR? Clearly, the White House could have intervened on behalf of the program: with EPA to speed up the approval of environmental permits, with OMB to resolve the Phase III controversy, and even with DOE to spur fast resolution of its personnel and procurement problems. But White House attention, especially to programmatic details, is a scarce resource. Programs with vocal and politically significant constituencies are much more likely to command White House attention than those without.

Although the SPR enjoys widespread intellectual support, it has no natural constituency. The SPR does not provide immediate benefits to narrowly defined groups; rather, it provides insurance against widely distributed economic losses that may occur sometime in the future. Defense programs provide benefits of a similar nature but they enjoy support from the sellers of weapon systems, veterans organizations, and a large and well established bureaucracy. Those who sold goods and services to the SPR program could sell them elsewhere and hence did not have much to gain from pushing for expansion. The SPR program did not enjoy the interest of organized groups; it was not highly visible to the general public. It was thrust into a disorganized bureaucracy with conflicting interests. With the exception of advocates of the RPR who were concerned with what they perceived to be the distribution of potential benefits, political action in support of the SPR was largely the result of the interest of individual congressmen. However, with the notable exception of forcing the Carter administration to resume filling the reserve, congressional oversight did not make a major contribution to implementation.

The absence of a politically significant constituency helps explain why the White House did not play an active role in reducing the bureaucratic impediments to implementation: other issues involving vocal constituencies claimed greater attention. Furthermore, as the political saliency of budget-cutting rose in the last two years of the Carter administration, the absence of a vocal constituency made the SPR a relatively attractive choice for expenditure reductions and delays.

Conclusion

Before being informed by my publisher that it might be misleading, I had intended to entitle this book *Wise Virgins* after the following parable from Matthew (Chapter 25, Verses 1-13):

Parable of the Ten Virgins

> Then will the kingdom of heaven be like ten virgins who took their lamps and went forth to meet the bridegroom and the bride. Five of them were foolish and five wise. But the five foolish, when they took their lamps, took no oil with the lamps. Then as the bridegroom was long in coming, they all became drowsy and slept. And at midnight a cry arose, "Behold, the bridegroom is coming, go forth to meet him!" Then all those virgins arose and trimmed their lamps. And the foolish said to the wise, "Give us some of your oil, for our lamps are going out." The wise answered, saying "Lest there may not be enough for us and for you, go rather to those who sell it, and buy some for yourselves."
>
> ... Watch therefore, for you know neither the day nor the hour.

A friend told me that his pastor once ended a sermon based on this parable with the following question: Will you stay awake with the wise virgins or sleep with the foolish? Although perhaps not with the same choice of phrase, we might ask a similar question about our preparedness for disruptions of the world oil market: Will our stocks of oil be adequate or will we be at the mercy of the foreign oil sellers?

Notes

1. David Howard Davis, *Energy Politics,* (New York: St Martin's Press, 1978).

2. Craufurd D. Goodwin, ed., *Energy Policy in Perspective: Today's Problems, Yesterday's Solutions* (Washington, D.C.: The Brookings Institution, 1981).

3. Thomas H. Tietenberg, *Energy, Planning and Policy: The Political Ecomony of Project Independence* (Lexington, Mass.: Lexington Press, 1976); Walter A. Rosenbaum, *Energy, Politics and Public Policy* (Washington, D.C.: Congressional Quarterly Press, 1981).

4. Aaron Wildavsky and Ellen Tenenbaum, *The Politics of Mistrust: Estimating American Oil and Gas Resources* (Beverly Hills, Calif.: Sage Publications, 1981).

5. A number of authors have speculated about the factors that are important in the life cycle of organizations. See, for example, Marver H. Bernstein, *Regulating Business by Independent Commission* (Princeton, N.J.: Princeton University Press, 1955); Anthony Downs, *Inside Bureaucracy* (Boston: Little, Brown and Company, 1966); and Herbert Kaufman, *Are Government Organizations Immortal*? (Washington, D.C.: The Brookings Institution, 1976). Of most relevance to our discussion is the growth of new organizations to maturity.

6. As an introduction to this literature, the reader might wish to refer to the following sample of recent works: Aaron Wildavsky, *Speaking Truth to Power: The Art and Craft of Policy Analysis* (Boston: Little, Brown and Company, 1979); Arnold J. Meltsner, *Policy Analysts in the Bureaucracy* (Berkeley: University of California Press, 1976); Henry J. Aaron, *Politics and the Professors* (Washington, D.C.: The Brookings Institution, 1978); Robert A. Goldwin, ed., *Bureaucrats, Policy Analysts, Statesmen: Who Leads?* (Washington, D.C.: American Enterprise Institute for Public Policy Research, 1980); Walter Williams, *Social Policy Research and Analysis* (New York: Elsevier, 1971); Robert Behn, "Policy Analysis and Policy Politics," *Policy Analysis* 7, no. 2 (spring 1981):199–226; James Q. Wilson, "Policy Intellectuals and Public Policy," *The Public Interest*, no. 64 (Summer

1981), pp. 31-46; and Michael J. Malbin, *Unelected Representatives: Congressional Staff and the Future of Representative Government* (New York: Basic Books, Inc., 1980).

7. Some examples include: Bruce Ackerman, Susan Rose-Ackerman, James W. Sawyer, Jr., and Dale W. Henderson, *The Uncertain Search for Environmental Quality* (New York: The Free Press, 1974); Gary D. Brewer, *Politicians, Bureaucrats, and the Consultant* (New York: Basic Books, 1973); and Martin Greenberger, Matthew A. Crenson, and Brian L. Crissey, *Models in the Policy Process* (New York: Russell Sage, 1976). Two volumes suggest regulatory decision making as a fertile ground for this line of inquiry: James C. Miller, III, and Bruce Yandle, eds., *Benefit-Cost Analysis of Social Regulation* (Washington D.C.: American Enterprise Institute for Public Policy Research, 1979); and Robert W. Crandall and Lester B. Lave, eds., *The Scientific Basis of Health and Safety Regulation* (Washington, D.C.: The Brookings Institution, 1981).

8. Hugh Heclo, "OMB and the Presidency—The Problem of Neutral Competence," *The Public Interest*, no. 38 (Winter 1975), pp. 80-98.

Appendix

The SPR Provisions of the Energy Policy and Conservation Act

Public Law 94-163

94th Congress, S. 622

December 22, 1975

PART B—STRATEGIC PETROLEUM RESERVE

DECLARATION OF POLICY

SEC. 151. (a) The Congress finds that the storage of substantial quantities of petroleum products will diminish the vulnerability of the United States to the effects of a severe energy supply interruption, and provide limited protection from the short-term consequences of interruptions in supplies of petroleum products.

42 USC 6231.

(b) It is hereby declared to be the policy of the United States to provide for the creation of a Strategic Petroleum Reserve for the storage of up to 1 billion barrels of petroleum products, but not less than 150 million barrels of petroleum products by the end of the 3-year period which begins on the date of enactment of this Act, for the purpose of reducing the impact of disruptions in supplies of petroleum products or to carry out obligations of the United States under the international energy program. It is further declared to be the policy of the United States to provide for the creation of an Early Storage Reserve, as part of the Reserve, for the purpose of providing limited protection from the impact of near-term disruptions in supplies of petroleum products or to carry out obligations of the United States under the international energy program.

DEFINITIONS

42 USC 6232.

SEC. 152. As used in this part:

(1) The term "Early Storage Reserve" means that portion of the Strategic Petroleum Reserve which consists of petroleum products stored pursuant to section 155.

(2) The term "importer" means any person who owns, at the first place of storage, any petroleum product imported into the United States.

(3) The term "Industrial Petroleum Reserve" means that portion of the Strategic Petroleum Reserve which consists of petroleum products owned by importers or refiners and acquired, stored, or maintained pursuant to section 156.

(4) The term "interest in land" means any ownership or possessory right with respect to real property, including ownership in fee, an easement, a leasehold, and any subsurface or mineral rights.

(5) The term "readily available inventories" means stocks and supplies of petroleum products which can be distributed or used without affecting the ability of the importer or refiner to operate at normal capacity; such term does not include minimum working inventories or other unavailable stocks.

(6) The term "refiner" means any person who owns, operates, or controls the operation of any refinery.

(7) The term "Regional Petroleum Reserve" means that portion of the Strategic Petroleum Reserve which consists of petroleum products stored pursuant to section 157.

(8) The term "related facility" means any necessary appurtenance to a storage facility, including pipelines, roadways, reservoirs, and salt brine lines.

(9) The term "Reserve" means the Strategic Petroleum Reserve.

(10) The term "storage facility" means any facility or geological formation which is capable of storing significant quantities of petroleum products.

(11) The term "Strategic Petroleum Reserve" means petroleum products stored in storage facilities pursuant to this part; such term includes the Industrial Petroleum Reserve, the Early Storage Reserve, and the Regional Petroleum Reserve.

STRATEGIC PETROLEUM RESERVE OFFICE

Establishment. 42 USC 6233.

SEC. 153. There is established, in the Federal Energy Administration, a Strategic Petroleum Reserve Office. The Administrator, acting through such Office and in accordance with this part, shall exercise authority over the establishment, management, and maintenance of the Reserve.

STRATEGIC PETROLEUM RESERVE

42 USC 6234.

SEC. 154. (a) A Strategic Petroleum Reserve for the storage of up to 1 billion barrels of petroleum products shall be created pursuant to this part. By the end of the 3-year period which begins on the date of enactment of this Act, the Strategic Petroleum Reserve (or the Early Storage Reserve authorized by section 155, if no Strategic Petroleum Reserve Plan has become effective pursuant to the provisions of section 159(a)) shall contain not less than 150 million barrels of petroleum products.

Plan to Congress. Post, p. 965.

(b) The Administrator, not later than December 15, 1976, shall prepare and transmit to the Congress, in accordance with section 551, a

Strategic Petroleum Reserve Plan. Such Plan shall comply with the provisions of this section and shall detail the Administrator's proposals for designing, constructing, and filling the storage and related facilities of the Reserve.

(c) (1) To the maximum extent practicable and except to the extent that any change in the storage schedule is justified pursuant to subsection (e)(6), the Strategic Petroleum Reserve Plan shall provide that:

(A) within 7 years after the date of enactment of this Act, the volume of crude oil stored in the Reserve shall equal the total volume of crude oil which was imported into the United States during the base period specified in paragraph (2);

(B) within 18 months after the date of enactment of this Act, the volume of crude oil stored in the Reserve shall equal not less than 10 percent of the goal specified in subparagraph (A);

(C) within 3 years after the date of enactment of this Act, the volume of crude oil stored in the Reserve shall equal not less than 25 percent of the goal specified in subparagraph (A); and

(D) within 5 years after the date of enactment of this Act, the volume of crude oil stored in the Reserve shall equal not less than 65 percent of the goal specified in subparagraph (A).

Volumes of crude oil initially stored in the Early Storage Reserve and volumes of crude oil stored in the Industrial Petroleum Reserve, and the Regional Petroleum Reserve shall be credited toward attainment of the storage goals specified in this subsection.

(2) The base period shall be the period of the 3 consecutive months, during the 24-month period preceding the date of enactment of this Act, in which average monthly import levels were the highest.

(d) The Strategic Petroleum Reserve Plan shall be designed to assure, to the maximum extent practicable, that the Reserve will minimize the impact of any interruption or reduction in imports of refined petroleum products and residual fuel oil in any region which the Administrator determines is, or is likely to become, dependent upon such imports for a substantial portion of the total energy requirements of such region. The Strategic Petroleum Reserve Plan shall be designed to assure, to the maximum extent practicable, that each noncontiguous area of the United States which does not have overland access to domestic crude oil production has its component of the Strategic Petroleum Reserve within its respective territory.

(e) The Strategic Petroleum Reserve Plan shall include:

(1) a comprehensive environmental assessment;

(2) a description of the type and proposed location of each storage facility (other than storage facilities of the Industrial Petroleum Reserve) proposed to be included in the Reserve;

(3) a statement as to the proximity of each such storage facility to related facilities;

(4) an estimate of the volumes and types of petroleum products proposed to be stored in each such storage facility;

(5) a projection as to the aggregate size of the Reserve, including a statement as to the most economically-efficient storage levels for each such storage facility;

(6) a justification for any changes, with respect to volumes or dates, proposed in the storage schedule specified in subsection (c), and a program schedule for overall development and completion of the Reserve (taking into account all relevant factors, including cost effectiveness, the need to construct related facilities, and the ability to obtain sufficient quantities of petroleum products to fill the storage facilities to the proposed storage levels);

(7) an estimate of the direct cost of the Reserve, including—
(A) the cost of storage facilities;
(B) the cost of the petroleum products to be stored;
(C) the cost of related facilities; and
(D) management and operation costs;

(8) an evaluation of the impact of developing the Reserve, taking into account—
(A) the availability and the price of supplies and equipment and the effect, if any, upon domestic production of acquiring such supplies and equipment for the Reserve;
(B) any fluctuations in world, and domestic, market prices for petroleum products which may result from the acquisition of substantial quantities of petroleum products for the Reserve;
(C) the extent to which such acquisition may support otherwise declining market prices for such products; and
(D) the extent to which such acquisition will affect competition in the petroleum industry;

(9) an identification of the ownership of each storage and related facility proposed to be included in the Reserve (other than storage and related facilities of the Industrial Petroleum Reserve);

(10) an identification of the ownership of the petroleum products to be stored in the Reserve in any case where such products are not owned by the United States;

(11) a statement of the manner in which the provisions of this part relating to the establishment of the Industrial Petroleum Reserve and the Regional Petroleum Reserve will be implemented; and

(12) a Distribution Plan setting forth the method of drawdown and distribution of the Reserve.

EARLY STORAGE RESERVE

42 USC 6235.

SEC. 155. (a)(1) The Administrator shall establish an Early Storage Reserve as part of the Strategic Petroleum Reserve. The Early Storage Reserve shall be designed to store petroleum products, to the maximum extent practicable, in existing storage capacity. Petroleum products stored in the Early Storage Reserve may be owned by the United States or may be owned by others and stored pursuant to section 156(b).

(2) If the Strategic Petroleum Reserve Plan has not become effective under section 159(a), the Early Storage Reserve shall contain not less than 150 million barrels of petroleum products by the end of the 3-year period which begins on the date of enactment of this Act.

(b) The Early Storage Reserve shall provide for meeting regional needs for residual fuel oil and refined petroleum products in any region which the Administrator determines is, or is likely to become, dependent upon imports of such oil or products for a substantial portion of the total energy requirements of such region.

Plan, transmittal to Congress.

(c) Within 90 days after the date of enactment of this Act, the Administrator shall prepare and transmit to the Congress an Early Storage Reserve Plan which shall provide for the storage of not less than 150 million barrels of petroleum products by the end of 3 years from the date of enactment of this Act. Such plan shall detail the Administrator's proposals for implementing the Early Storage Reserve requirements of this section. The Early Storage Reserve Plan shall, to the maximum extent practicable, provide for, and set forth the manner in which, Early Storage Reserve facilities will be incorporated into the Strategic Petroleum Reserve after the Strategic Petroleum Reserve Plan has become effective under section 159(a). The Early Storage Reserve Plan shall include, with respect to the Early Storage Reserve, the same or similar assessments, statements, estimates, evaluations, projections, and other information which section 154(c) requires to be included in the Strategic Petroleum Reserve Plan, including a Distribution Plan for the Early Storage Reserve.

INDUSTRIAL PETROLEUM RESERVE

Establishment. 42 USC 6236.

SEC. 156. (a) The Administrator may establish an Industrial Petroleum Reserve as part of the Strategic Petroleum Reserve.

(b) To implement the Early Storage Reserve Plan or the Strategic Petroleum Reserve Plan which has taken effect pursuant to section 159(a), the Administrator may require each importer of petroleum products and each refiner to (1) acquire, and (2) store and maintain in readily available inventories, petroleum products in amounts determined by the Administrator, except that the Administrator may not require any such importer or refiner to store such petroleum products in an amount greater than 3 percent of the amount imported or refined by such person, as the case may be, during the previous calendar year. Petroleum products imported and stored in the Industrial Petroleum Reserve shall be exempt from any tariff or import license fee.

(c) The Administrator shall implement this section in a manner which is appropriate to the maintenance of an economically sound and competitive petroleum industry. The Administrator shall take such steps as are necessary to avoid inequitable economic impacts on refiners and importers, and he may grant relief to any refiner or importer who would otherwise incur special hardship, inequity, or unfair distribution of burdens as the result of any rule, regulation, or order promulgated under this section. Such relief may include full or partial exemption from any such rule, regulation, or order and the issuance of an order permitting such an importer or refiner to store petroleum products owned by such importer or refiner in surplus storage capacity owned by the United States.

REGIONAL PETROLEUM RESERVE

42 USC 6237.

SEC. 157. (a) The Strategic Petroleum Reserve Plan shall provide for the establishment and maintenance of a Regional Petroleum Reserve in, or readily accessible to, each Federal Energy Administration Region, as defined in title 10, Code of Federal Regulations in effect on November 1, 1975, in which imports of residual fuel oil or any refined petroleum product, during the 24-month period preceding the date of computation, equal more than 20 percent of demand for such oil or product in such regions during such period, as determined by the Administrator. Such volume shall be computed annually.

(b) To implement the Strategic Petroleum Reserve Plan, the Administrator shall accumulate and maintain in or near any such Federal Energy Administration Region described in subsection (a), a Regional Petroleum Reserve containing volumes of such oil or product, described in subsection (a), at a level adequate to provide substantial protection against an interruption or reduction in imports of such oil or product to such region, except that the level of any such Regional Petroleum Reserve shall not exceed the aggregate volume of imports of such oil or product into such region during the period of the 3 consecutive months, during the 24-month period specified in subsection (a), in which average monthly import levels were the highest, as determined by the Administrator. Such volume shall be computed annually.

(c) The Administrator may place in storage crude oil, residual fuel oil, or any refined petroleum product in substitution for all or part of the volume of residual fuel oil or any refined petroleum product stored in any Regional Petroleum Reserve pursuant to the provisions of this section if he finds that such substitution (1) is necessary or desirable for purposes of economy, efficiency, or for other reasons, and (2) may be made without delaying or otherwise adversely affecting the fulfillment of the purpose of the Regional Petroleum Reserve.

OTHER STORAGE RESERVES

Report to Congress.
42 USC 6238.

SEC. 158. Within 6 months after the Strategic Petroleum Reserve Plan is transmitted to the Congress, pursuant to the requirements of section 154(b), the Administrator shall prepare and transmit to the Congress a report setting forth his recommendations concerning the necessity for, and feasibility of, establishing—

(1) Utility Reserves containing coal, residual fuel oil, and refined petroleum products, to be established and maintained by major fossil-fuel-fired baseload electric power generating stations;

(2) Coal Reserves to consist of (A) federally-owned coal which is mined by or for the United States from Federal lands, and (B) Federal lands from which coal could be produced with minimum delay; and

(3) Remote Crude Oil and Natural Gas Reserves consisting of crude oil and natural gas to be acquired and stored by the United States, in place, pursuant to a contract or other agreement or arrangement entered into between the United States and persons who discovered such oil or gas in remote areas.

REVIEW BY CONGRESS AND IMPLEMENTATION

42 USC 6239.

SEC. 159. (a) The Strategic Petroleum Reserve Plan shall not become effective and may not be implemented, unless—

(1) the Administrator has transmitted such Plan to the Congress pursuant to section 154(b); and

(2) neither House of Congress has disapproved (or both Houses have approved) such Plan, in accordance with the procedures specified in section 551.

(b) For purposes of congressional review of the Strategic Petroleum Reserve Plan under subsection (a), the 5 calendar days described in section 551(f)(4)(A) shall be lengthened to 15 calendar days, and the 15 calendar days described in section 551 (c) and (d) shall be lengthened to 45 calendar days.

Post, p. 965.

(c) The Administrator may, prior to transmittal of the Strategic Petroleum Reserve Plan, prepare and transmit to the Congress proposals for designing, constructing, and filling storage or related facilities. Any such proposal shall be accompanied by a statement explaining (1) the need for action on such proposals prior to completion of such Plan, (2) the anticipated role of the proposed storage or related facilities in such Plan, and (3) to the maximum extent practicable, the same or similar assessments, statements, estimates, evaluations, projections, and other information which section 154(c) requires to be included in the Strategic Petroleum Reserve Plan.

(d) The Administrator may prepare amendments to the Strategic Petroleum Reserve Plan or to the Early Storage Reserve Plan. He shall transmit any such amendment to the Congress together with a statement explaining the need for such amendment and, to the maximum extent practicable, the same or similar assessments, statements, estimates, evaluations, projections, and other information which section 154(e) requires to be included in the Strategic Petroleum Reserve Plan.

(e) Any proposal transmitted under subsection (c) and any amendment transmitted under subsection (d), other than a technical or clerical amendment or an amendment to the Early Storage Reserve Plan, shall not become effective and may not be implemented unless—

(1) the Administrator has transmitted such proposal or amendment to the Congress in accordance with subsection (c) or (d) (as the case may be), and

(2) neither House of Congress has disapproved (or both Houses of Congress have approved) such proposal or amendment, in accordance with the procedures specified in section 551.

(f) To the extent necessary or appropriate to implement—

(1) the Strategic Petroleum Reserve Plan which has taken effect pursuant to subsection (a);

(2) the Early Storage Reserve Plan;

(3) any proposal described in subsection (c), or any amendment described in subsection (d), which such proposal or amendment has taken effect pursuant to subsection (e); and

(4) any technical or clerical amendment or any amendment to the Early Storage Reserve Plan,

the Administrator may:

(A) promulgate rules, regulations, or orders;

(B) acquire by purchase, condemnation, or otherwise, land or interests in land for the location of storage and related facilities;

(C) construct, purchase, lease, or otherwise acquire storage and related facilities;

(D) use, lease, maintain, sell, or otherwise dispose of storage and related facilities acquired pursuant to this part;

(E) acquire, subject to the provisions of section 160, by purchase, exchange, or otherwise, petroleum products for storage in the Strategic Petroleum Reserve, including the Early Storage Reserve and the Regional Petroleum Reserve;

(F) store petroleum products in storage facilities owned and controlled by the United States or in storage facilities owned by others if such facilities are subject to audit by the United States;

(G) execute any contracts necessary to carry out the provisions of such Strategic Petroleum Reserve Plan, Early Storage Reserve Plan, proposal or amendment;

(H) require any importer of petroleum products or any refiner to (A) acquire, and (B) store and maintain in readily available inventories, petroleum products in the Industrial Petroleum Reserve, pursuant to section 156;

(I) require the storage of petroleum products in the Industrial Petroleum Reserve, pursuant to section 156, on such reasonable terms as the Administrator may specify in storage facilities owned and controlled by the United States or in storage facilities other than those owned by the United States if such facilities are subject to audit by the United States;

(J) require the maintenance of the Industrial Petroleum Reserve;

(K) maintain the Reserve; and

(L) bring an action, whenever he deems it necessary to implement the Strategic Petroleum Reserve Plan, in any court having jurisdiction of such proceedings, to acquire by condemnation any real or personal property, including facilities, temporary use of facilities, or other interests in land, together with any personal property located thereon or used therewith.

(g) Before any condemnation proceedings are instituted, an effort shall be made to acquire the property involved by negotiation, unless, the effort to acquire such property by negotiation would, in the judgment of the Administrator be futile or so time-consuming as to unreasonably delay the implementation of the Strategic Petroleum Reserve Plan, because of (1) reasonable doubt as to the identity of the owners, (2) the large number of persons with whom it would be necessary to negotiate, or (3) other reasons.

PETROLEUM PRODUCTS FOR STORAGE IN THE RESERVE

42 USC 6240.

SEC. 160. (a) The Administrator is authorized, for purposes of implementing the Strategic Petroleum Reserve Plan or the Early Storage Reserve Plan, to place in storage, transport, or exchange—

(1) crude oil produced from Federal lands, including crude oil produced from the Naval Petroleum Reserves to the extent that such production is authorized by law;

(2) crude oil which the United States is entitled to receive in kind as royalties from production on Federal lands; and

(3) petroleum products acquired by purchase, exchange, or otherwise.

(b) The Administrator shall, to the greatest extent practicable, acquire petroleum products for the Reserve, including the Early Storage Reserve and the Regional Petroleum Reserve in a manner consonant with the following objectives:

(1) minimization of the cost of the Reserve;

(2) orderly development of the Naval Petroleum Reserves to the extent authorized by law;

(3) minimization of the Nation's vulnerability to a severe energy supply interruption;

(4) minimization of the impact of such acquisition upon supply levels and market forces; and

(5) encouragement of competition in the petroleum industry.

DRAWDOWN AND DISTRIBUTION OF THE RESERVE

42 USC 6241.

SEC. 161. (a) The Administrator may drawdown and distribute the Reserve only in accordance with the provisions of this section.

(b) Except as provided in subsections (c) and (f), no drawdown and distribution of the Reserve may be made except in accordance with the provisions of the Distribution Plan contained in the Strategic Petroleum Reserve Plan which has taken effect pursuant to section 159(a).

(c) Drawdown and distribution of the Early Storage Reserve may be made in accordance with the provisions of the Distribution Plan contained in the Early Storage Reserve Plan until the Strategic Petroleum Reserve Plan has taken effect pursuant to section 159(a).

(d) Neither the Distribution Plan contained in the Strategic Petroleum Reserve Plan nor the Distribution Plan contained in the Early Storage Reserve Plan may be implemented, and no drawdown and distribution of the Reserve or the Early Storage Reserve may be made, unless the President has found that implementation of either such Distribution Plan is required by a severe energy supply interruption or by obligations of the United States under the international energy program.

(e) The Administrator may, by rule, provide for the allocation of any petroleum product withdrawn from the Strategic Petroleum Reserve in amounts specified in (or determined in a manner prescribed by) and at prices specified in (or determined in a manner prescribed by) such rules. Such price levels and allocation procedures shall be consistent with the attainment, to the maximum extent practicable, of the objectives specified in section 4(b) (1) of the Emergency Petroleum Allocation Act of 1973.

Rules.

15 USC 753.

(f) The Administrator may permit any importer or refiner who owns any petroleum products stored in the Industrial Petroleum Reserve pursuant to section 156 to remove or otherwise dispose of such products upon such terms and conditions as the Administrator may prescribe.

COORDINATION WITH IMPORT QUOTA SYSTEM

SEC. 162. No quantitative restriction on the importation of any petroleum product into the United States imposed by law shall apply to volumes of any such petroleum product imported into the United States for storage in the Reserve.

42 USC 6242.

DISCLOSURE, INSPECTION, INVESTIGATION

SEC. 163. (a) The Administrator may require any person to prepare and maintain such records or accounts as the Administrator, by rule, determines necessary to carry out the purposes of this part.

Record-keeping.
42 USC 6243.

(b) The Administrator may audit the operations of any storage facility in which any petroleum product is stored or required to be stored pursuant to the provisions of this part.

(c) The Administrator may require access to, and the right to inspect and examine, at reasonable times, (1) any records or accounts required to be prepared or maintained pursuant to subsection (a) and (2) any storage facilities subject to audit by the United States under the authority of this part.

NAVAL PETROLEUM RESERVES STUDY

SEC. 164. The Administrator shall, in cooperation and consultation with the Secretary of the Navy and the Secretary of the Interior, develop and submit to the Congress within 180 days after the date of enactment of this Act, a written report recommending procedures for the exploration, development, and production of Naval Petroleum Reserve Number 4. Such report shall include recommendations for protecting the economic, social, and environmental interests of Alaska Natives residing within the Naval Petroleum Reserve Number 4 and analyses of arrangements which provide for (1) participation by private industry and private capital, and (2) leasing to private industry. The Secretary of the Navy and the Secretary of the Interior shall cooperate fully with one another and with the Administrator: the Secretary of the Navy shall provide to the Administrator and Secretary of the Interior all relevant data on Naval Petroleum Reserve Number 4 in order to assist the Administrator in the preparation of such report.

Report to Congress.
42 USC 6244.

ANNUAL REPORTS

SEC. 165. The Administrator shall report to the President and the Congress, not later than one year after the transmittal of the Strategic Petroleum Reserve Plan to the Congress and each year thereafter, on all actions taken to implement this part. Such report shall include—

Report to Congress and President.
42 USC 6245.

(1) a detailed statement of the status of the Strategic Petroleum Reserve;

(2) a summary of the actions taken to develop and implement the Strategic Petroleum Reserve Plan and the Early Storage Reserve Plan;

(3) an analysis of the impact and effectiveness of such actions on the vulnerability of the United States to interruption in supplies of petroleum products;

(4) a summary of existing problems with respect to further implementation of the Early Storage Reserve Plan and the Strategic Petroleum Reserve Plan; and

(5) any recommendations for supplemental legislation deemed necessary or appropriate by the Administrator to implement the provisions of this part.

AUTHORIZATION OF APPROPRIATIONS

42 USC 6246.

SEC. 166. There are authorized to be appropriated—

(1) such funds as are necessary to develop and implement the Early Storage Reserve Plan (including planning, administration, acquisition, and construction of storage and related facilities) and as are necessary to permit the acquisition of petroleum products for storage in the Early Storage Reserve or, if the Strategic Petroleum Reserve Plan has become effective under section 159(a), for storage in the Strategic Petroleum Reserve in the minimum volume specified in section 154(a) or 155(a)(2), whichever is applicable; and

(2) $1,100,000,000 to remain available until expended to carry out the provisions of this part to develop the Strategic Petroleum Reserve Plan and to implement such plan which has taken effect pursuant to section 159(a), including planning, administration, and acquisition and construction of storage and related facilities, but no funds are authorized to be appropriated under this paragraph for the purchase of petroleum products for storage in the Strategic Petroleum Reserve.

Bibliography

Aaron, Henry J. *Politics and the Professors* (Washington, D.C.: The Brookings Institution, 1978).

Ackerman, Bruce A., Susan Rose-Ackerman, James W. Sawyer, Jr., and Dale W. Henderson. *The Uncertain Search for Environmental Quality* (New York: The Free Press, 1974).

Adelman, M. A. "Oil Insurance." *Washington Post*, July 7, 1980, p. A-15.

The Aerospace Corporation. "An Overview of the Strategic and Critical Materials Stockpiling Program and Its Relationship to SPR Sizing." Prepared for the Strategic Petroleum Reserve Office, U.S. Department of Energy. September 1980.

The Aerospace Corporation. "SPR Size Studies Review," Draft Final Report. Prepared for the Strategic Petroleum Reserve Office, U.S. Department of Energy. November 15, 1980.

Anthony, Robert N. *Planning and Control Systems: A Framework for Analysis* (Cambridge, Mass.: Harvard University Press, 1965).

Asher, Norman J., and Wendy West. "Use of Tankers for Stockpiling Petroleum," Institute for Defense Analyses, Paper P-1241, January, 1977.

Bachman, W. A. "Problems Plague U.S. Crude Storage Program," *Oil and Gas Journal,* August 6, 1979, pp. 49-53.

Balas, Egan. "The Strategic Petroleum Reserve: How Large Should It Be? Management Science Research Report No. 436, Graduate School of

Industrial Administration, Carnegie-Mellon University, June 30, 1979.

Bardach, Eugene. *The Implementation Game: What Happens After a Bill Becomes a Law* (Cambridge, Mass.: MIT Press, 1977).

Barron, Michael. "Market-oriented Financing of Oil Stockpile Acquisition," Staff Working Paper, Office of Policy, Planning and Analysis, Department of Energy, February 15, 1981.

Behn, Bobert. "Policy Analysis and Policy Politics," *Policy Analysis* 7, no. 2 (Spring 1981):199-226.

Berman, Larry. *The Office of Management and Budget and the Presidency 1921-1979* (Princeton, N.J.: Princeton University Press, 1979).

Bernstein, Marver H. *Regulating Business by Independent Commission* (Princeton, N.J.: Princeton Unviersity Press, 1955).

Blankenship, Jerry, Mike Barron, Joseph Eschbach, Linsay Bower, and William Lane. "The Energy Problem: Costs and Policy Options," U.S. Department of Energy, Assistant Secretary for Policy and Evaluation, Office of Oil, May 23, 1980.

Bohi, Douglas R., and Milton Russell. *U.S. Energy Policy: Alternatives for Security* (Baltimore: Johns Hopkins University Press, 1975), pp. 94-99.

Brewer, Gary D. *Politicians, Bureaucrats, and the Consultant.* (New York: Basic Books, 1973).

Childress, J. Philip. "Description: Strategic Reserve Cost/Benefit Model," FEA, Office of the Assistant Administrator for Policy and Analysis, August 20, 1975.

Childress, J. Philip, Glenn Coplan, and Lowell Goodhue. "Strategic Storage Program Preliminary Cost/Benefit Analysis," Draft, Federal Energy Administration, August 29, 1975.

Choa, Hung-Po, and Alan S. Manne. "Oil Stockpiles and Import Reductions: A Dynamic Programming Approach," Electric Power Institute, Palo Alto, California, October 1980.

Comptroller General of the U.S. "Capability of the Naval Petroleum and Oil Shale Reserves to Meet Emergency Oil needs," GAO, October 5, 1972, B-66927.

———. "Factors Influencing the Size of the U.S. Strategic Petroleum Reserve," GAO, June 15, 1979, ID-79-8.

———. "Information on Department of Energy's Management of the Strategic Petroleum Reserve," GAO, March 22, 1979, EMD-79-49.

———. "Issues Needing Attention in Developing the Strategic Petroleum Reserve," GAO, February 16, 1977, EMD-77-20.

———. "Questionable Suitability of Certain Salt Caverns and Mines for the Strategic Petroleum Reserve," GAO, August 14, 1978, EMD-75-65.

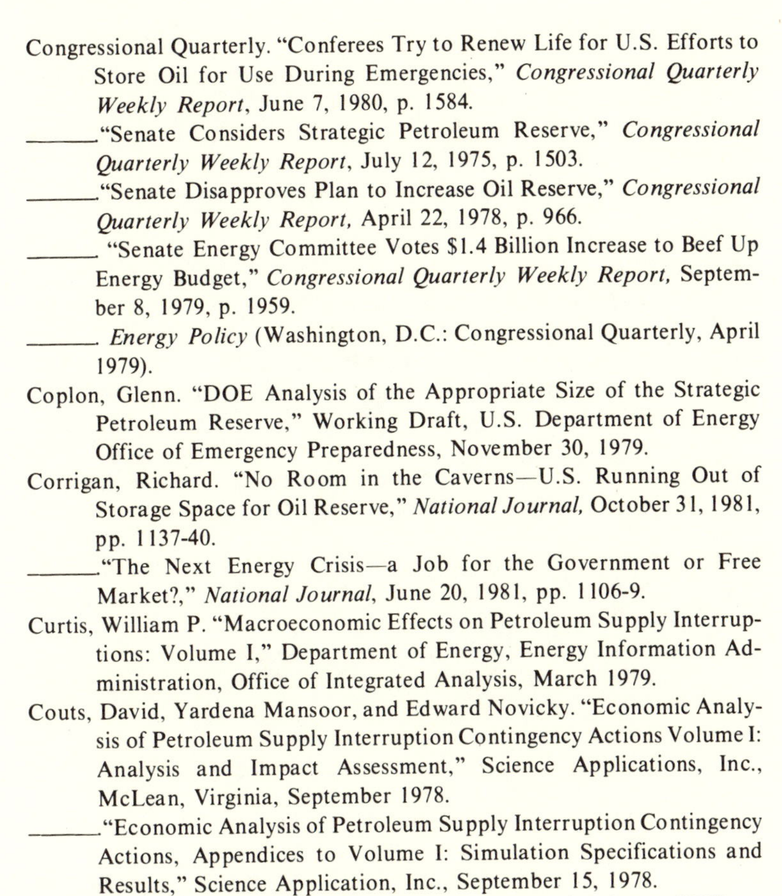

Congressional Quarterly. "Conferees Try to Renew Life for U.S. Efforts to Store Oil for Use During Emergencies," *Congressional Quarterly Weekly Report*, June 7, 1980, p. 1584.

______."Senate Considers Strategic Petroleum Reserve," *Congressional Quarterly Weekly Report*, July 12, 1975, p. 1503.

______."Senate Disapproves Plan to Increase Oil Reserve," *Congressional Quarterly Weekly Report*, April 22, 1978, p. 966.

______. "Senate Energy Committee Votes $1.4 Billion Increase to Beef Up Energy Budget," *Congressional Quarterly Weekly Report*, September 8, 1979, p. 1959.

______. *Energy Policy* (Washington, D.C.: Congressional Quarterly, April 1979).

Coplon, Glenn. "DOE Analysis of the Appropriate Size of the Strategic Petroleum Reserve," Working Draft, U.S. Department of Energy Office of Emergency Preparedness, November 30, 1979.

Corrigan, Richard. "No Room in the Caverns—U.S. Running Out of Storage Space for Oil Reserve," *National Journal*, October 31, 1981, pp. 1137-40.

______."The Next Energy Crisis—a Job for the Government or Free Market?," *National Journal*, June 20, 1981, pp. 1106-9.

Curtis, William P. "Macroeconomic Effects on Petroleum Supply Interruptions: Volume I," Department of Energy, Energy Information Administration, Office of Integrated Analysis, March 1979.

Couts, David, Yardena Mansoor, and Edward Novicky. "Economic Analysis of Petroleum Supply Interruption Contingency Actions Volume I: Analysis and Impact Assessment," Science Applications, Inc., McLean, Virginia, September 1978.

______."Economic Analysis of Petroleum Supply Interruption Contingency Actions, Appendices to Volume I: Simulation Specifications and Results," Science Application, Inc., September 15, 1978.

Davis, David H. *Energy Politics* (New York: St. Martin's Press, 1978).

Deese, David A., and Joseph S. Nye, eds. *Energy and Security* (Cambridge, Mass: Ballinger Publishing Company, 1981).

Devarajan, Shantayanen, and Anthony C. Fisher. "Hotelling's 'Economics of Exhaustible Resources': Fifty Years Later," *Journal of Economic Literature* 19, no. 1 (March 1981): 65-73.

Dohner, Robert S., "Energy Prices, Economic Activity and Inflation: A Survey of the Issues," presented at the Conference on Energy Prices, Inflation, and Economic Activity, MIT Center for Energy Policy Research, November 7-9, 1979.

Downs, Anthony, *Inside Bureaucracy* (Boston: Little, Brown and Company, 1966).

Economic Regulatory Administration, "Notice of Proposed Rulemaking: Distribution of Strategic Petroleum Reserve Crude Oil," *Federal Register,* August 20, 1979, pp. 48696-707.

Executive Office of the President. "The National Energy Plan," April 29, 1977 (Washington, D.C.: Government Printing Office, 1977).

Federal Energy Administration *Project Independence Report,* November 1974.

———. Office of International Affairs. "U.S. Oil Companies and the Arab Oil Embargo: The International Allocation of Constrained Supplies," prepared for the Subcommittee on Multinational Corporations of the Committee on Foreign Relations, U.S. Senate, January 27, 1975 (Washington, D.C.: Government Printing Office, 1975).

The Federation of American Scientists. "Special Issue: Strategic Petroleum Reserve," *Journal of the Federation of American Scientists* 33, no. 9 (November 1980).

Felton, John. "Congress Clears $11.6 Billion Bill for Interior, Energy," *Congressional Quarterly Weekly Report*, October 14, 1978, pp. 2929-31.

Goldwin, Robert A., ed. *Bureaucrats, Policy Analysts, Statesmen: Who Leads?* (Washington, D.C.: American Enterprise Institute for Public Policy Research, 1980).

Goodwin, Craufurd D., ed. *Energy Policy in Perspective: Today's Problems, Yesterday's Solutions* (Washington, D.C.: The Brookings Institution, 1981).

Gramlich, Edward M. "Macro Policy Responses to Price Shocks," in *Brookings Papers on Economic Activity*, edited by Arthur Okun and George L. Perry, no. 1 (1979), pp. 125–66.

Greenberger, Martin, Matthew A. Crenson, and Brian L. Crissey. *Models in the Policy Process* (New York: Russell Sage, 1976).

Harberger, Arnold C. "Three Basic Postulates of Welfare Economics: An Interpretative Essay," *Journal of Economic Literature* 9, no. 3 (September 1971): 785-97.

Haveman, Joel, and James G. Phillips. "Energy Report/Independence Blueprint Weighs Various Options," *National Journal Reports*, November 2, 1974, pp. 1635–54.

Heal, Geoffrey, "The Relationship Between Price and Extraction Cost of a Resource With a Backstop Technology," *Bell Journal of Economics* 7, no. 2. (Autumn 1976): 371–78.

Heclo, Hugh. "OMB and the Presidency—The Problem of Neutral Competence." *Public Interest*, no. 38 (Winter 1975): 80–98.

Hillier, Frederick S., and Gerald J. Lieberman, *Introduction to Operations Research* (San Francisco: Holden-Day, Inc. 1972).

Hogan, William, W., "Oil Stockpiling: Help Thy Neighbor," Energy and Environmental Policy Center, John F. Kennedy School of Government, Harvard University, Cambridge, Massachusetts, March 1982.

Holcombe, Randall G. "A Method for Estimating the GNP Loss from a Future Oil Embargo," *Policy Sciences* 8, no. 1 (June 1977): 217–34.

Holt, Barry J. and Mark Berkman, "An Evaluation of the Strategic Petroleum Reserve," U.S. Congress, Congressional Budget Office, Natural Resources and Commerce Division, June 1980.

Horwich, George. "Government Contigency Planning for Petroleum Supply Interruptions: A Macro Perspective," presented at the Conference on Policies for Coping with Oil Supply Disruptions, American Enterprise Institute for Public Policy Research, Washington, D.C., September 8-9, 1980.

Hotelling, Harold. "The Economics of Exhaustible Resources," *The Journal of Political Economy* 39, no. 2 (April 1931): 137–75.

Hystad, Carlyle E., "Estimating Appropriate Reserve Size: 750 and 1000 MMB," FEA, Strategic Petroleum Reserve Office, March 24, 1977.

Jennrich, John H. "SPR Became a Political Football," *Oil and Gas Journal*, June 2, 1980, p. 65.

———. "In-Kind Crude Tariff Pushed for SPR," *Oil and Gas Journal*, December 15, 1980, p. 46.

Jippe, Bob and Richard Wheatley, "High Cost of Bulging Inventories Compounds Problems for U.S. Refiners," *Oil and Gas Journal*, June 22, 1981, pp. 19-22.

Johany, Ali D., "OPEC and the Price of Oil: Cartelization or Alteration of Property Rights?", *Journal of Energy Development*. 5, no. 1. (Autumn 1979), pp. 72-80.

JRB Associates, Inc., "Feasibility Study for Requiring Storage of Crude Oil, Residual Fuel Oil, and/or Refined Petroleum Products by Industry," Final Report to the Federal Energy Administration, December 2, 1976.

———. "Feasibility Study for Requiring Storage of Crude Oil, Residual Fuel Oil and/or Selected Petroleum Products by Industry: Amendment No. 1 to Final Report," January 15, 1977.

Kaufman, Herbert. *Are the Government Organizations Immortal?* (Washington, D.C.: The Brookings Institution, 1976).

Kinberg, Yerman, Melvin F. Shakun, and Ephriam F. Sudit. "Energy Buffer Stock Decisions in Game Situations," in *Energy Policy*, edited by J. S. Aronofsky, A. G. Rao, and M. D. Shokun (New York: North-Holland Publishing Company, 1978), pp. 109–27.

Klass, Michael W., James C. Burrows, and Steven D. Beggs. *International Minerals Cartels and Embargoes* (New York: Praeger, 1980).

Krapels, Edward N. "Focus On Emergency Oil Reserves," *Petroleum Economist* 48, no. 2 (February 1981): 46-48.

______. *Oil Crisis Management: Strategic Stockpiling for International Security* (Baltimore: Johns Hopkins University Press, 1980).

Kuenne, Robert E., Gerald F. Higgins, Robert J. Michaels, and Mary Summerfield. "A Policy to Protect the U.S. Against Oil Embargoes," *Policy Analysis* 1, no. 4 (Fall 1975): 571-97.

Kuenne, Robert, E., Jerry Blankenship, and Paul F. McCoy. "Optimal Drawdown Strategy for Strategic Petroleum Reserves," Institute for Defense Analyses, April 1977.

______. "Optimal Drawdown Patterns for Strategic Petroleum Reserves," *Energy Economics* 1, no. 1 (January 1979): 3-13.

Landsberg, Hans H., ed. *Energy the Next Twenty Years* (Cambridge, Mass.: Ballinger Publishing Company, 1979).

Lawson, Robert G. "Strategic Petroleum Reserve Construction Ends First Phase," *Oil and Gas Journal*, July 21, 1980, pp. 47-53.

Leuba, H. R. "A Free Enterprise Oil Storage Corporation," Jack Faucett Associates, Chevy Chase, Md., July 1980.

Madison, Christopher. "How Can We Build an Oil Reserve Without Offending the Saudis?" *National Journal*, June 28, 1980, pp. 1044-49.

______. "The Energy Department at Three—Still Trying to Establish Itself," *National Journal*, November 4, 1980, pp. 1644-49.

Malbin, Michael J., *Unelected Representatives: Congressional Staff and the Future of Representative Government* (New York: Basic Books, Inc., 1980).

Marver, James D., *Consultants Can Help* (Lexington, Mass.: Lexington Books, 1979).

Mead, Walter J. "An Economic Analysis of Crude Oil Price Behavior in the 1970's," *Journal of Energy and Development* 4, no. 2 (Spring 1979): 212-28.

______, and Phillip E. Sorensen. "A National Defense Petroleum Reserve Alternative To Oil Import Quotas," *Land Economics* 47, no. 3 (August 1971): 211-24.

Meltsner, Arnold J. *Policy Analysts in the Bureaucracy* (Berkeley: University of California Press, 1976), pp. 48-49.

Miller, James C., III, and Bruce Yandle, eds., *Benefit-Cost Analysis of Social Regulation* (Washington, D.C.: American Enterprise Institute for Public Policy Research, 1979).

Mitchell, Edward J. "Protection for Petroleum Refiners," *Regulation*, July/August 1981, pp. 37-42.

Mork, Knut, and Robert Hall. "Macroeconomic Analysis of Energy Price Shocks and Offsetting Policies: An Integrated Approach," in

Energy Prices, Inflation, and Economic Activity, edited by K. A. Mork (Cambridge, Mass.: Ballinger Publishing Company, 1981).

National Petroleum Council. "Emergency Preparedness for Interruptions of Petroleum Imports into the United States," Proposed Final Interim Report, July 24, 1973.

______. "Emergency Preparedness for Interruptions of Petroleum Imports into the United States," April 1981.

______. "Petroleum Storage for National Security," August 1975.

National Petroleum Council's Committee on Emergency Preparedness. *Emergency Preparedness for Interruption of Petroleum Imports into the United States* (Washington, D.C.: National Petroleum Council (September 1974).

Nichols, A. L., and R. J. Zeckhauser. "Stockpiling Strategies and Cartel Prices," *Bell Journal of Economics* 8, no. 1 (1977):66-96.

Novicky, Edward R. "A Review of Analytical Techniques and Studies Related to Assessing the Impacts of Petroleum Shortfalls," STSC, Management Technology Division, Bethesda, Md., May 1979.

Nordhaus, William D. "The Energy Crisis and Macroeconomic Policy," *The Energy Journal* 1, no. 1 (January 1980):11-20.

______. "The 1974 Report of the President's Council of Economic Advisors: Energy in the Economic Report," *American Economic Review* 64, no. 4 (September 1974):556-65.

Oil and Gas Journal. "DOE Slates 100,000 b/d of oil for SPR by Dec. 1," *Oil and Gas Journal*, September 22, 1980, p. 57.

Oil and Gas Journal. "Senate Moves to Speed NPR-A Drilling, SPR Fill," *Oil and Gas Journal*, November 24, 1980, p. 66.

Paulson, S. Lawrence. "Administration Undecided on Strategic Petroleum Reserve Size, Panel is Told," *The Oil Daily*, September 16, 1980, p. 2.

Pauly, David, and William J. Cook, "Salting Away a Reserve," *Newsweek*, August 25, 1980.

Pelham, Ann. "Energy Department Trying to Work Out Problems of Costly Storage Program," *Congressional Quarterly Weekly Report*, February 3, 1979, pp. 204-5.

Peterson, Frederick M. and Anthony C. Fisher, "The Exploitation of Extractive Resources: A Survey," *Economic Journal*, 87 (December 1977), pp. 681-721.

Pindyck, Robert S. "Energy Price Increases and Macroeconomic Policy," *The Energy Journal* 1, no. 4 (October 1980):1-20.

______. "Gains to Producers from the Cartelization of Exhaustible Resources," *Review of Economics and Statistics* 60, no. 2 (May 1978): 238-51.

______. "The War, and Oil Prices," *New York Times*, December 9, 1980, p. A27.

Plummer, James L. "Methods for Measuring the Oil Import Reduction Premium and the Oil Stockpile Premium," *The Energy Journal* 2, no. 1 (January 1981):1-18.

Pressman, Jeffrey L., and Aaron Wildavsky. *Implementation* (Berkeley: University of California Press, 1975).

Raiffa, Howard, *Decision Analysis* (Reading, Mass.: Addison-Wesley, 1968).

Ridgeway, James. "New Ideas to Move Oil Reserve Purchases Off Budget," *The Energy Daily* 8, no. 226 (December 1980):1, 3.

Rosenbaum, Walter A. *Energy, Politics and Public Policy* (Washington, D.C.: Congressional Quarterly Press, 1981).

Rowen, Henry, and John Weyant. "The Optimal Strategic Petroleum Reserve Size for the U.S.?" Stanford University, International Energy Program Discussion Paper, October 1979.

Sapolsky, Harvey M. *The Polaris Systems Development* (Cambridge, Mass.: Harvard University Press, 1972).

Simon, Julian L., "Global Confusion 1980: A hard Look at the Global 2000 Report," *The Public Interest*, no. 62 (Winter 1981), pp. 3-20.

Snyder, Glenn H. *Stockpiling Strategic Materials: Politics and National Defense* (San Franciso: Chandler Publishing Company, 1966).

Sobotka and Company, Inc. "Option for Placing Current Excess Private Crude Oil Stocks in the SPR," draft, Washington, D. C., August 29, 1980.

Soloman, Burt, "Strength in Numbers: Independent Refiners Seek Own Access to Crude," *Energy Daily*, March 30, 1981.

Strategic Petroleum Reserve Office.* "Annual Report," February 16, 1980, DOE/RA-0047.

———. "Annual Report," February 16, 1978, DOE/RA-0004/1(77).

———. "Annual Strategic Petroleum Reserve Report," February 16, 1979, DOE/US-0003.

———. "Early Storage Plan," April 22, 1976.

———. "Impacts of Regulation on the Strategic Petroleum Reserve: A Selected Analysis," October 15, 1979.

———. "Other Storage Reserves Report," August 16, 1977.

———. "Program Stewardship Report No. 1: SPR Baseline," November 3, 1978.

———. "Program Stewardship Report No. 2: SPR Status and Issues," January 25, 1979.

———. Strategic Petroleum Reserve Annual Report, February 16, 1981, DOE/RA-0047/1.

*All material published by the Strategic Petroleum Reserve is now available through the Department of Energy, Washington, D. C.

———. "Strategic Petroleum Reserve Plan," December 15, 1976.

———. "Stategic Petroleum Reserve Plan Amendment No. 1: Acceleration of the Development Schedule," *Energy Action*, no. 12, February 16, 1977.

———. "Strategic Petroleum Reserve Plan Amendment No. 2 (Energy Action DOE No. 1): Expansion of the Strategic Petroleum Reserve," March 1978, DOE/RA-0032/2.

———. "Strategic Petroleum Reserve Plan Amendment No. 3 (Energy Action DOE No. 5): Distribution Plan for the Strategic Petroleum Reserve," DOE/RA/039, October 1979.

Sweetnam, Glen. "Reducing the Costs of Oil Interruptions: The Role of the Strategic Petroleum Reserve," presented at the ORSA/TIMS Joint National Meeting, Colorado Springs, Colorado, November 10, 1980.

Sweetnam, Glen, George Horwich, and Steve Minihan. "An Analysis of Acquisition and Drawdown Strategies for the Strategic Petroleum Reserve," Draft, U. S. Department of Energy, Assistant Secretary for Policy and Evaluation, Office of Oil, December 17, 1979.

Symonds, Edward. "Stockpile Programme Begins Again—But Long Way to Go," *Petroleum Economist* 47, no. 10 (October 1980):417–18.

Tani, Steven N., and Dean W. Boyd. "Measuring the Economic Cost of an Oil Embargo," Stanford Research Institute, Stanford, Calif., October 1976.

Teisberg, Thomas J. "A Dynamic Programming Model of the U. S. Strategic Petroleum Reserve," *Bell Journal of Economics* 12, no. 2 (Autumn 1981):526–46.

Tolley, G. S., and J. D. Wilman. "The Foreign Dependence Question," *The Journal of Political Economy* 85, no. 2 (April 1977):323–47.

Tietenberg, Thomas H., *Energy Planning and Policy: The Political Economy of Project Independence* (Lexington, Mass.: Lexington Books, 1976).

Tokyo Communique. "Joint Declaration of Tokyo Summit Conference," June 29, 1979. In *Public Papers of the Presidents of the United States, Jimmy Carter, Book II, June 23 to December 31, 1979* (Washington, D.C.: Government Printing Office, 1980), pp. 1197–1201.

U. S. Army Corps of Engineers. "Management Plan Development of Environmental Impact Statements for Turnkey Sites," Huntsville, Alabama, August 20, 1979.

U. S. Congress, Congressional Budget Office. "An Evaluation of the Strategic Petroleum Reserve," June 1980 (Washington, D.C.: Government Printing Office, 1980).

———. "Energy Policy and Conservation Act," Public Law 94–163, December 22, 1975, 89 Stat. 871.

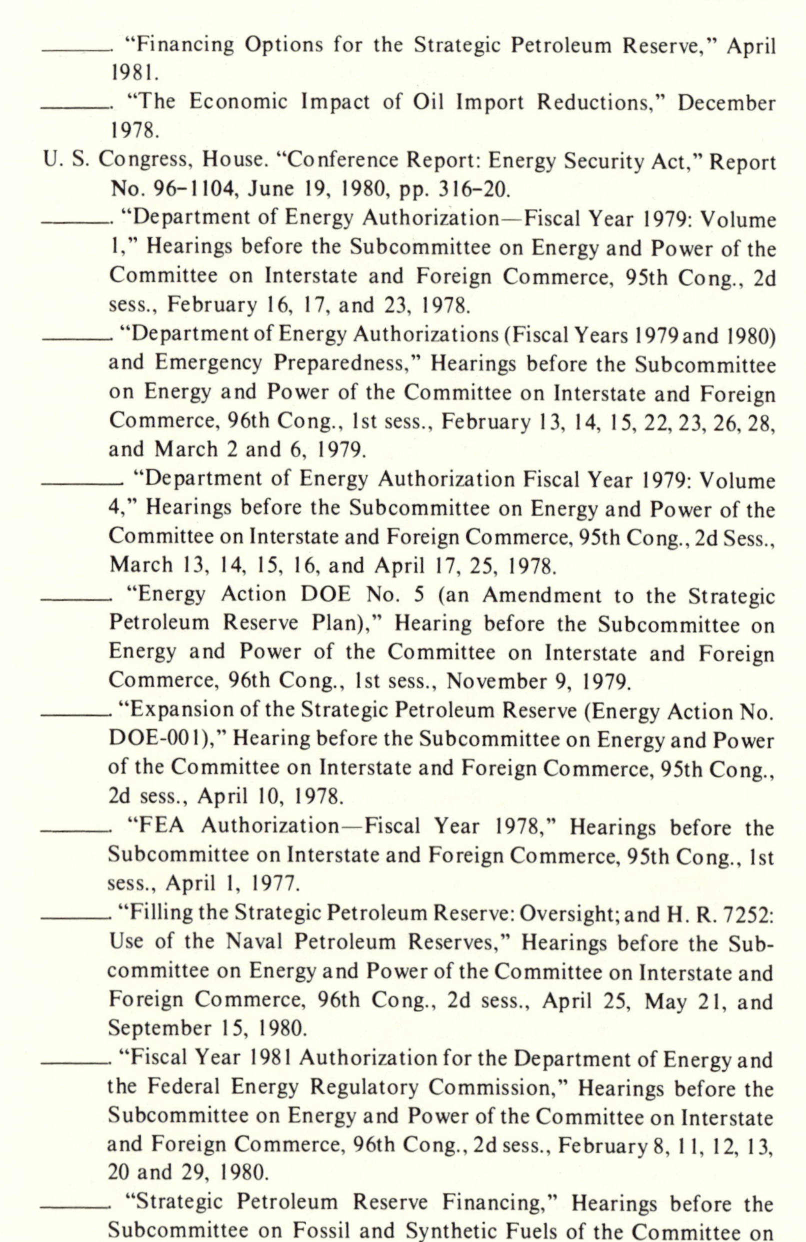

———. "Financing Options for the Strategic Petroleum Reserve," April 1981.

———. "The Economic Impact of Oil Import Reductions," December 1978.

U. S. Congress, House. "Conference Report: Energy Security Act," Report No. 96–1104, June 19, 1980, pp. 316–20.

———. "Department of Energy Authorization—Fiscal Year 1979: Volume 1," Hearings before the Subcommittee on Energy and Power of the Committee on Interstate and Foreign Commerce, 95th Cong., 2d sess., February 16, 17, and 23, 1978.

———. "Department of Energy Authorizations (Fiscal Years 1979 and 1980) and Emergency Preparedness," Hearings before the Subcommittee on Energy and Power of the Committee on Interstate and Foreign Commerce, 96th Cong., 1st sess., February 13, 14, 15, 22, 23, 26, 28, and March 2 and 6, 1979.

———. "Department of Energy Authorization Fiscal Year 1979: Volume 4," Hearings before the Subcommittee on Energy and Power of the Committee on Interstate and Foreign Commerce, 95th Cong., 2d Sess., March 13, 14, 15, 16, and April 17, 25, 1978.

———. "Energy Action DOE No. 5 (an Amendment to the Strategic Petroleum Reserve Plan)," Hearing before the Subcommittee on Energy and Power of the Committee on Interstate and Foreign Commerce, 96th Cong., 1st sess., November 9, 1979.

———. "Expansion of the Strategic Petroleum Reserve (Energy Action No. DOE-001)," Hearing before the Subcommittee on Energy and Power of the Committee on Interstate and Foreign Commerce, 95th Cong., 2d sess., April 10, 1978.

———. "FEA Authorization—Fiscal Year 1978," Hearings before the Subcommittee on Interstate and Foreign Commerce, 95th Cong., 1st sess., April 1, 1977.

———. "Filling the Strategic Petroleum Reserve: Oversight; and H. R. 7252: Use of the Naval Petroleum Reserves," Hearings before the Subcommittee on Energy and Power of the Committee on Interstate and Foreign Commerce, 96th Cong., 2d sess., April 25, May 21, and September 15, 1980.

———. "Fiscal Year 1981 Authorization for the Department of Energy and the Federal Energy Regulatory Commission," Hearings before the Subcommittee on Energy and Power of the Committee on Interstate and Foreign Commerce, 96th Cong., 2d sess., February 8, 11, 12, 13, 20 and 29, 1980.

———. "Strategic Petroleum Reserve Financing," Hearings before the Subcommittee on Fossil and Synthetic Fuels of the Committee on

Energy and Commerce, 97th Cong., 1st sess., March 18, April 28, and May 11, 1981.

______. "Strategic Petroleum Reserves," Hearing before the Subcommittee on Energy and Power of the Committee on Interstate and Foreign Commerce, 95th Cong., 1st sess., February 16, 1977.

______. "Strategic Petroleum Reserves: Oil Supply and Construction Problems," Hearing before the Subcommittee on Energy and Power of the Committee on Interstate and Foreign Commerce, 96th Cong., 1st sess., September 10, 1979.

______. "Strategic Petroleum Reserves: Reprogramming of Funds," Hearing before the Subcommittee on Energy and Power of the Committee on Interstate and Foreign Commerce, 95th Cong., 2d sess., December 18, 1978.

U. S. Congress, Senate. "Amendment No. 1 to the Strategic Petroleum Reserve," Hearing before the Subcommittee on Energy and Natural Resources, 95th Cong., 1st sess., June 9, 1977.

______. "Oversight of the Structure and Management of the Department of Energy," Staff Report, Committee on Governmental Affairs, 96th Cong., 2d sess., December 1980.

______. "Strategic Petroleum Reserve—DOE Energy Action No. 1," Hearing before the Subcommittee on Energy Production and Supply of the Committee on Energy and Natural Resources, 95th Cong., 2d sess., April 10, 1978.

______. "Strategic Petroleum Reserves," Hearings before the Committee on Interior and Insular Affairs, 93d Cong., 1st sess., May 30 and July 26, 1973.

______. "Disapproving Energy Action DOE No. 1," Report No. 95-738, April 13, 1978.

______. "Review of the Strategic Petroleum Reserve Plan," Hearing before the Committee on Interior and Insular Affairs, 95th Cong., 1st sess., February 4, 1977.

______. Committee on Governmental Affairs. "Oversight of the Structure and Management of the Department of Energy," Staff Report, December 1980 (Washington, D.C.: Government Printing Office, 1980).

U.S. Department of Energy. "Communication from the Secretary of Energy Transmitting Energy Action No. DOE-002, to Expand the Size of the Strategic Petroleum Reserve to 1 Billion Barrels, Pursuant to Section 159(d) of the Energy Policy and Conservation Act of 1975 (Public Law 94-163)." House Document No. 95-339, May 19, 1978 (Washington, D. C.: Government Printing Office, 1978).

______. "Macroeconomic Effects of Petroleum Supply Interruptions: Volumes I and II," March 1979 (Washington D. C.: Government Printing Office, 1979).

———. "National Energy Plan II: A Report to the Congress Required by Title VIII of the Department of Energy Organization Act (Public Law 95-91)," May 1979.

———. "Report on the Explosion, Fire, and Oil Spill Resulting in One Fatality and Injury on September 21, 1978, at Well 6 of Cavern 6 at the West Hackberry, Louisiana, Oil Storage Site of the Strategic Petroleum Reserve" DOE/EV-0032. November 1978 (Washington, D.C.: Government Printing Office, 1978).

———, Assistant Secretary for Policy and Evaluation. "Reducing U.S. Oil Vulnerability: Energy Policy for the 1980s," November 10, 1980 (Washington, D.C.: Government Printing Office, 1980).

———, Office of Policy, Planning and Analysis, Energy Security Office, "An Emergency Preparedness Strategy for Oil Supply Disruptions," Draft, May 30, 1981.

———. SPR Files, "Draft Budget Issue Paper," RA Volume IVA1-A40, B1-B21, March 21, 1980.

———, Strategic Petroleum Reserve Task Force, "DOE Analysis of the Appropriate Size of the Strategic Petroleum Reserve," Draft, October 11, 1979.

———. "DOE Analysis of the Need for the Fourth 250 MMB of the Strategic Petroleum Reserve and DOE Comments on OMB's Position," December 18, 1978.

———, Task Force on Regional and Noncontiguous Storage, "Program/Project Plan for Regional and Noncontiguous Petroleum Reserve Storage Project," Preliminary Report, May 30, 1980.

Varian, Hal R. *Microeconomic Analysis.* (New York: W. W. Norton and Company, 1978).

Verleger, Philip K., Jr. "Let the Market Fill the U. S. Petroleum Reserve," *Wall Street Journal*, April 27, 1981, editorial page.

———. "The U. S. Petroleum Crisis of 1979," in *Brookings Papers on Economic Activity*, edited by Arthur Okun and George L. Perry, no. 2 (1979), pp. 463–76.

Vielvoye, Roger. "The Futures Market," *Oil and Gas Journal*, February 16, 1981, p. 66.

Weimer, David L. "Routine SPR Acquisitions: The In-Kind Import Tariff and Spot Market Purchase Authority," Office of Oil Staff Paper, Policy and Evaluation Office, Department of Energy, September 10, 1980.

Wildavsky, Aaron. *Speaking Truth to Power: The Art and Craft of Policy Analysis* (Boston: Little, Brown and Company, 1979).

———, and Ellen Tenenbaum. *The Politics of Mistrust: Estimating American Oil and Gas Resources* (Beverly Hills, California: Sage Publications, 1981).

Williams, Walter. *Social Policy Research and Analysis* (New York: Elsevier, 1971).

Willig, Robert D. "Consumer's Surplus Without Apology," *American Economic Review* 65, no. 4. (September 1976):589-97.

Wilson, James Q. "Policy Intellectuals and Public Policy," *The Public Interest*, no. 64 (Summer 1981):31-46.

Index

Acquisition, 31-32; controversy, 64-72; optimal, 77, 125; spot price mandate, 77; strategies for, 76-77; world price impact of, 66
Akins, James E., on vulnerability, 10
Analysis, in bureaucratic battles, 178
Arab-Israeli conflict, 5
Arab oil embargo, xi, 11

Barging, 41
Basic sales agreements (BSA), 171
Bradley, William, 70
Brill, Jay R.: on DOE bureaucracy, 54; as SPR Office director, 57
Brine disposal, feasibility studies, 46-47
BSA. *See* Basic sales agreements
Bureaucratic environment: expertise in, 40; and formal analysis, 106, 193-96; and methodological innovation, 106; and neutral competence role, 194; policy entrepreneurs in, 135; program implementation, 16-18, 59-61, 191-93; project execution, 17-18; for stockpiling decisions, 103-7; White House role in, 195-96. *See also* Strategic Petroleum Reserve Office

Camp David Accords, and market disruptions, 112
Carter administration: acquisition decisions, 63; price decontrol, 166-67; and regional petroleum reserves, 168-69
Civil service system, 192-93
Consultants, as expertise source, 40
Consumers' surplus: compensating variation, 95; equivalent variation, 108 n.9; measurement, 92-93
Contingency planning, interagency task force on, 118
Cost/schedule tradeoff, 40-42
Crude oil. *See* Oil

Davies, Robert L., 53; on size issue, 114; and strategic planning, 22-23
Davis, David Howard, *Energy Politics*, 188-89
Decision analysis: numerical example of, 101 (figure); for risk accounting, 99-102; for stockpiling problems, 100 (figure)
Defense Fuel Supply Center (DFSC), 64; SPR procurement, 31-32
DeLuca, Joseph R., as SPR Office director, 56-57, 149-54
Department of Energy (DOE): analytic debate with OMB, 119-20, 194; creation of, 43, 53-55, 190; Emergency Response Planning Office, 170; issue importance within, 135; and SPR credibility, 154
Dependence, distinguished from vulnerability, 4, 190
Deutch, John: on SPR purchases, 65; on turnkey program, 152-53
Dingell, John: on SPR fill decisions, 70; on turnkey program, 154
Disruption scenarios, comparisons of, 128-29
Distribution, 170; auction approach, 173-74, 179; market determined, 170; regulatory approach, 171-73
DOE. *See* Department of Energy
Dole, Robert, on SPR fill decisions, 70
Drawdowns: capability for, 32; formal modeling of, 177-79; GNP benefits of, 98; and IEA, 181; leakage problem, 123; military benefits, 84; policy formulation, 164-65, 188; revenue benefits, 123; strategies for, 125, 174-79; trigger mechanisms, 175-77
Duncan, Charles W., 65-66
Dynamic programming, 107 n.2; in decision analysis, 102-3; for drawdown problem, 177-79; and optimal stockpile size, 125

Early Storage Reserve (ESR), 23-24
Econometric models, 96-97
Economic losses, without SPR, 133
Economic Regulatory Administration (ERA), 171-72
Edwards, Edwin W., 149; and DOE concessions, 50-51
EIS. *See* Environmental impact statements
Emergency Petroleum Allocation Act of 1973 (PL93-159), 12, 170
Eminent domain powers, 50
Energy policy: national security aspects of, 116; SPR relation to, 188. *See also* Vulnerability
Energy Policy and Conservation Act of 1975 (EPCA), 141, 170; drawdown requirements, 174-75; established SPR Office, 22, 114; IPR authority, 143-44; passage of, 13-15; petroleum reserves policy of, 113; regional reserve requirement 165; strategic planning, 191
Energy Politics (Davis), 189
Energy Research and Development Administration (ERDA), 43. *See also* Department of Energy
Energy Security Act (PL96-294), SPR amendment, 70-71
Energy value: price theory, 189; quantity theory, 189-91
Entitlements system: as import subsidy, 5; and SPR purchases, 31

Environmental impact statements (EIS): for ESR plan, 24; programmatic, 50
Environmental Protection Agency (EPA): on brine disposal, 47, 192; permitting process, 52
Environmental regulations, 62 n.5
EPA. *See* Environmental Protection Agency
EPCA. *See* Energy Policy and Conservation Act
Equity financing. *See* financing
ERA. *See* Economic Regulatory Administration
ERDA. *See* Energy Research and Development Administration
Erdoelbevorratungsverband (EBV), 156-58
Exhaustible resources, optimal extraction rate, 87

FEA. *See* Federal Energy Administration
Federal budget, 112
Federal Energy Administration (FEA): and Project Independence, 12; special procurement board, 42-43; SPR advantages under, 190
Financing, 72-76; alternatives, 187; equity, 72-73, 74, 159, 188; in-kind import fee, 75-76; off-budget, 72
Ford, Gerald, and Energy Independence Act of 1975, 13
Foreign governments, oil stockpiling in, 85
Foreign policy options, 84

General Accounting Office (GAO), 35
General equilibrium analysis, 92-93
GNP. *See* Gross national product
Goodwin, Craufurd D., 189
Gross national product: measurement, 109; relation to social welfare, 96

Helco, Hugh, 194
Hotelling, Harold, on exhaustible resources, 87
Hystad, Carlyle, 158; as acting SPR director, 53-54; and procurement problem, 43-44

IEA. *See* International Energy Agency
Imported oil: elasticity of demand for, 139 n.35; mandatory quotas, 9; reduction policies, 124; retaliatory cutbacks in, 84-85; social costs of, 4, 76, 187; voluntary restrictions on, 9. *See also* Oil; Tariff
Industrial petroleum reserve (IPR): authority for, 14, 141; and budget outlays, 155; feasibility study, 144-45; incentives for, 85; low visibility of purchases, 148; as OMB alternative, 112; options, 156; political expediency of, 26; private sector opposition to, 145; regulatory complexity of, 147; taking issue, 146-47; tax subsidies, 157
Inflation, reduction benefits, 96
Input/output model, 97
Institute for Defense Analyses, 114
International Energy Agency (IEA): sharing agreements, 123, 124; U.S. obligations to, 174
International Energy Program Agreement, 35, 141

International monetary system, 95
IPR. *See* Industrial petroleum reserve
Iran: crude oil production, 64, 68, 148; revolution in, xi; and SPR expansion, 153

Jackson, Henry M., 10-11
Jones, Harry A., as SPR Office director, 57

Kassebaum, Nancy A., 74-75
Kennedy, Edward M., 36-37, 168

Louisiana Department of Natural Resources, 51-52

McClure, James A., 74
McDonald, Stephen, 9
Macroeconomic models: demand-driven, 129; disruption simulations, 120-21, 128; and size issue, 118-20
Management control, 17, 41-42
Marine Engineers' Beneficial Association, 36
Markey, Edward, 169
Mead, Walter, 9
Middle East: political stability of, xi, 5, 97; war possibility, 84
Moffett, Toby, 169

National Energy Plan (NEP), 111, 117
National Environmental Policy Act of 1969 (PL91-190), 50
National Fuels and Energy Policy Study, 10
National Oceanic and Atmospheric Administration (NOAA), 52
National Petroleum Council, 113-14
National Security Council, 119
National Wildlife Federation, 35
Natural gas, decontrol of, 13
Naval petroleum reserve (NPR): for filling SPR, 15, 65; operational readiness, 8; as surge capacity, 7-8; trading for SPR oil, 71
Naval Petroleum Reserves Production Act of 1976 (PL94-258), 7
Neutral competence. *See* Bureaucratic environment
New England Council, 166
Nixon, Richard M., 12
Noel, Thomas E., 25, 45, 53
NPR. *See* Naval petroleum reserve
Nye, Joseph, 158

OAPEC. *See* Organization of Arab Petroleum Exporting Countries
Office of Management and Budget: analytic debate with DOE, 122-23, 134, 194; joint benefit study with DOE, 119-20; Phase IV objections of, 119; and SPR funding, 55; SPR hiring freeze, 53; SPR implementation delay, 112; and SPR size, 117
Office of Natural Gas and Integrated Analysis, 124-25
Oil: consumption reductions, 133; in critical materials program, 7; derived demand for, 93; future prices of, 86; futures markets, 176; in-kind import fee, 75-76; marginal social cost of, 108 n.8; market structure, 86, 153; price assumptions, 89 (figure); price controls, 5, 143, 167-68, 188; price drag, 94; price elasticity of demand, 124; price shocks, 73, 94,

95 (figure); property rights shift, 87; in real price trends, 64, 87-88; regional markets, 167; residual fuel market, 166; state prorationing, 6; stocks accumulation, 65, 68; storage capacity, 29; substitutes, 87-88; and supply disruption, 89-90; surge production, 6; type specification, 30-31; world market conditions, 3-4, 64, 86, 190. *See also* Imported oil; Royalty oil
OPEC. *See* Organization of Petroleum Exporting Countries
Organizational environment. *See* Bureaucratic environment
Organization of Arab Petroleum Exporting Countries (OAPEC), 11
Organization of Petroleum Exporting Countries (OPEC): as oligopoly, 86-87; production reductions, 66-67

Partial equilibrium analysis, 93-94
Petroleum Industry Research Foundation, 36
Petroleum Reserves and Import Policy Act of 1973, 10-11, 143
Politics of Mistrust, The (Tenenbaum; Wildavsky), 189
Presidential authority, on drawdowns, 175
Private Equity Petroleum Reserve Act (H.R. 2304), 73
Private stockpiles: externalities, 142-43; incentives for, 155; of risk-neutral firms, 165 n.2; speculation, 143; strategic reserves conversion, 159-61. *See also* Stockpiles
Program implementation. *See* Bureaucratic environment
Project Independence, 12-13; cost-effective reserve size, 114; report, 190
Project Independence Evaluation System (PIES), 115

Quantitative economic analysis: shortcomings of, 103-6; SPR role of, 194

Reagan administration: and equity financing, 74; oil purchase decisions, 63-64; price theory of energy value, 191
Regional petroleum reserves (RPR), 14-15, 191; controversy over, 32-33, 165-70; funding elimination for, 170; as price regulation complement, 168; and SPR Plan, 36-37, 146
Rosenbaum, Walter A., 189
Rotterdam, storage lease in, 56
Royalty oil: for SPR, 31; and SPR costs, 35. *See also* Oil
RPR. *See* Regional Petroleum Reserve

Saudi Arabia: and Arab/Israeli conflict resolution, 69; 1980 production decisions, 69; oil reserves, 68; and SPR filling, 79 n.22, 154; threatened production reductions, 67-68
Scenario/break-even methodology, 98-99
Schlesinger, James R., 56; as energy advisor, 44-45; and OMB, 113, 119; resignation, 135; on RPR, 168; on SPR acceleration, 45; and

SPR size, 116-17; on turnkey program, 150
Sell-option, 160-61
Social benefits: macroeconomic estimation, 94-97; microeconomic estimation, 92-94; relation to GNP, 96; scenario/break-even methodology, 98-99; surplus measure, 127; and uncertainty, 98
Social costs: measurement of, 91-92
Solution mining, 27-29; expertise shortages, 46
Sorensen, Philip, 9
SPR. *See* Strategic petroleum reserve
Stockman, David: on SPR fill decisions, 70; as SPR supporter, 130
Stockpiles: accumulation cost of, 89; analysis assumptions, 104-5 (table); carrying costs, xi; consequences of, 86-90; costs/benefits of, 90-97; drawdown benefits, 84, 90; expected price for, 142; in situ storage, 8, 19 n.8; optimal size for, 83; as policy, 6-9, 185-88; price effects of purchases, 88-89; storage methods, 8; strategic materials program, 6; as supply disruption deterrence, 84; value of, 3. *See also* Private stockpiles; Strategic petroleum reserve
Strategic and Critical Materials Stock Piling Revision Act of 1979 (PL96-41)
Strategic petroleum reserve (SPR): acceleration decision, 44-46; appropriations for, 58 (table); as asset, 72; budget impact, 31; capacity, 47; constituency, xii, 195-96; and construction contracting, 42; cost overruns, 112; credibility of, 122; current status, 57-59; enabling legislation for, 13-14; game theoretic model for, 107 n.1; implementation, 15-18, 33-35, 34 (figure), 39; as insurance, 76, 133; net economic benefits, 132 (table); noneconomic benefits, 85, 131-33; organizational problems, 18; Phase I facilities, 48-49 (table); private sector proposals, 154-61; and private stocks transfers, 66; for private storage, 158-59; and producer retaliation, 148; program failures, 40-56; salvage value of, 123; site selection, 29-30; size issue, 14, 25-26, 131-36; storage alternatives, 23; storage development, 187; two reserve plan, 121-24. *See also* Acquisition; Drawdowns; Financing; Regional petroleum reserves; Stockpiles
Strategic Petroleum Reserve Amendments of 1981 (S. 707), 74
Strategic Petroleum Reserve Office: capabilities of, 45; credibility of, 135; educational burden on, 54; environmental permits, 50-53; established, 43; FEA position of, 24-25; local politics, 50-53; location issue, 27, 44; management issue, 55, 56-57, 135; new acquisition system, 71-72; organization of, 186-87; personnel problems, 53, 196. *See also* Bureaucratic environment
Strategic Petroleum Reserve Plan: congressional approval, 35-38; contract management, 191-92; distribution amendment, 171; drawdowns, 175; IPR option in, 143-48; 1978 amendment to, 117-19; and optimal size estimates, 115, 116; Phase III, 124-31, 134,

187; Phase IV, 134, 187; planning inadequacies of, 46-50; RPR question in, 165-66; storage alternatives in, 28 (table); trigger mechanisms, 181 n.19
Strategic planning, 16; major issues, 25-35; process overview, 22-25
Supply disruption: consequences of, 86; GNP-loss function, 123-24; macroeconomic costs of, 118
Synthetic Fuels Corporation, 70
Systems design, 16-17

Taking issue, 146-47
Tariff, 76; efficiency of, 4; to finance SPR, 187-88; optimal, 78 n.29
Teisberg, Thomas J. 102, 136; dynamic programming model, 125-26
Teisberg model: CEA critique of, 126-27; loss function for, 129; OMB objections to, 127-28; on Phase III question, 130; on storage capacity, 131
Tenenbaum, Ellen, *The Politics of Mistrust*, 189
Tietenberg, Thomas H., 12, 189
Tokyo Economic Summit, 65
Trade Agreements Extension Act of 1955 (PL84-86), 9
Turnkey program, 149-54, 162 n.14; cancellation of, 153, 189; near-term budget costs, 153; solicitation, 151

Uncertainty: in optimal strategies models, 125; in stockpiles evaluation, 97-103
Unemployment, reduction benefits, 96
United States: Middle East policy, 5; as oil exporter, 7
United States Army Corps of Engineers: permitting process, 52; turnkey projects management, 150

Vulnerability: distinguished from dependence, 4, 190; to oil market disruptions, 186; reduction methods, 190

Warner, John W., 74
West Germany, IPR requirements of, 154
Wildavsky, Aaron, *The Politics of Mistrust*, 189
Windfall Profits Tax, 169

Zarb, Frank G., 14, 24-25

About the Author

DAVID LEO WEIMER holds degrees in engineering, public policy, and statistics, and is an Assistant Professor of Political Science and Public Policy Analysis at the University of Rochester. He previously served as a Visiting Economist at the Office of Policy, Planning and Analysis of the U.S. Department of Energy. Professor Weimer is the author of *Improving Prosecution* (Greenwood Press, 1980), and articles on criminal justice policy and the politics of regulation which have appeared in *Policy Sciences*, *Public Choice*, *Policy Analysis*, *Public Administration Review*, and the *Journal of Criminal Justice*.